RANDOM PROCESSES IN PHYSICAL SYSTEMS

RANDOM PROCESSES IN PHYSICAL SYSTEMS

AN INTRODUCTION TO PROBABILITY-BASED COMPUTER SIMULATIONS

CHARLES A. WHITNEY

A WILEY-INTERSCIENCE PUBLICATION

John Wiley & Sons, Inc.

NEW YORK / CHICHESTER / BRISBANE / TORONTO / SINGAPORE

Library of Congress Cataloging in Publication Data:

Whitney, Charles Allen.
    Random processes in physical systems: an introduction to probability-based
computer simulations/Charles A. Whitney.
        p.   cm.
    "A Wiley-Interscience publication."
    Includes bibliographical references.
    ISBN 0-471-51792-5
    1. Science—Computer simulation.   2. Random walks (Mathematics)
3. Programming (Electronic computers)   I. Title.
Q183.9.W49   1990   89-25111
501'.13—dc20   CIP

Printed in the United States of America

10 9 8 7 6 5 4 3 2 1

*Dedicated with gratitude
to
students and staff
of
The Harvard Freshman Seminar Program*

PREFACE

For surely not by planning did prime bodies find rank and place,
  nor by intelligence,
nor did they regulate movement by sworn pact,
but myriad atoms sped such myriad ways
  from the All forever, pounded,
    pushed, propelled,
by weight of their own launched and speeding along,
joining all possible ways,
  trying all forms,
whatever their meeting in congress could create;
and thus it happens that, widespread down the ages,
attempting junctures and movements of all kinds,
they at last formed patterns which,
  when joined together,
became at once the origin of great things,
earth,
    sea,
      and sky,
          and life in all its forms....

The Nature of Things, by Lucretius: Book V, 1, 419–431. (Frank O. Copley, translator, New York: Norton, 1977)

The advent of powerful desktop computers promises to open a new door to the natural sciences and mathematics. It is often called "numerical experimentation" and it consists of computer programs that permit the manipulation of simulated physical systems. This is not a substitute for true laboratory simulation, but it is inexpensive and powerful, and it can take us into areas that were formerly off-bounds. By controlling these simulations, we can explore our misconceptions and reconstruct our understanding. We can move into new areas of chemistry, physics, and statistics. We can learn to apply new techniques to simple problems and the gradually extend them to more complex situations—with no more mathematics than intermediate algebra.

This book introduces the reader to computer methods that are modeled on the rules of gambling. They can be done on a personal computer, and they can provide physical and mathematical insights to the expert and novice alike.

My intent is to discuss physical properties of gases and plasmas from a quantitative, but *physical*, rather than a mathematical, point of view. I feel that most introductions to statistical physics and thermodynamics are so formalized and so steeped in the classical assumptions of equilibrium theory that they have little bearing on practical problems. As an astrophysicist, I have found that my most interesting research has taken me into areas where equilibrium assumptions are guilty until proven innocent, rather than the other way around. In the atmosphere of a star, for example, it is possible to define a half-dozen different types of temperature. In such cases, the idea of temperature should be abandoned altogether, and with the aid of computer simulations this has become feasible.

So my goal is to encourage among my readers a more pragmatic and critical attitude toward the application of physical concepts than is implied by the immediate assumption of classical methods.

A formalist will notice that in many instances I have used a concept or term before defining it. This is intentional, as I believe that this is the natural way to acquire new vocabulary. I have attempted to make this book self-contained, so it might be used as a tutorial for a motivated student. Or it might be used to provide structure for an undergraduate seminar course or as a supplement to a conventional course in probability or elementary physics. Advanced students and researchers will find new approaches to familiar problems, and some of the simulations in this book will challenge sophisticated understanding, while at the same time being accessible to undergraduates and advanced high school students.

During the past decade, I have taught a course for college freshmen based on the material treated in this book, and I have weeded and sorted the topics in the search for those that seemed most accessible and powerful. The process of the seminar consisted primarily of

(a) weekly group discussion in which I posed problems and demonstrated short computer simulations for analysis,
(b) weekly assignments of simulations intended as bridges between group discussions, and
(c) individual projects that occupied the second half of the semester and were described by the students during a final "show-and-tell" session.

Numerous exercises are scattered throughout the text. They were written as an integral part of the book, and they contain many results not explicitly mentioned elsewhere. I have attempted to phrase them so that the hurried reader can glean some of their meaning without actually doing the labor.

Each chapter is focused on a particular simulation, and I describe some of the key points in the programming. The student in encouraged to construct at least one program for each chapter and use it as an exploratory tool. As hints, I provide flowcharts and program fragments, occasionally in *Basic* although primarily in *Pascal*. For the reader who has a computer but prefers to avoid high-level programming, a spreadsheet can provide a rapid and easy way to obtain numerical results for many problems (see Appendix 3). They are particularly efficient for seeing the effect of changing numerical parameters, and they also provide virtually instantaneous tabular or graphical output with a minimum of programming. For this reason, they have been used to produce many of the results illustrated in this book.

The prologue is an essay on the ideas underlying the modern approach to randomness. The body of the text is divided into four parts. Part 1 introduces pseudo-random numbers and answers question such as, "How can a computer generate randomness?" and "Why do we find patterns in randomness?" The "random walk" is the principal theme, as well as the excitation of atoms whose energy arrives in discrete quanta. This part provides an introduction to the modern descriptions of gases known as kinetic theory and thermodynamics. Part 2 describes a few of the most important processes that take place in the continuum of time, especially the scattering of photons in a gas and the "Brownian motion" of small particles. Part 3 applies these modeling techniques to the behavior of more complex systems and points the way to what I expect to be a major use of computers in the future. Part 4 introduces the application of randomizing methods to the solution of statistical problems, such as curve-fitting and error analysis.

The first two appendixes contain elementary introductions to the rules and terminology of probability and statistics, and they provide definitions and examples for many of the terms used in the text. The third appendix introduces the use of spreadsheet programs to problems in physics and statistics. The advantages of this approach are ease of programming and the automatic generation of graphs. A few literature references are given at the end of each chapter so the interested reader can explore further, and the complete bibliography is at the end of the book.

CHARLES A. WHITNEY

CONTENTS

PROLOGUE

P.1 SCIENCE AS THE SEARCH FOR WHAT NEEDS TO BE EXPLAINED

Modern scientists seek to *explain*, and a key step is to decide what needs explaining and what does not. Looking about us, we see shapes—a tree, a snowflake, a constellation in the night sky—and we wonder how they arose. Some of them are mere accidents, not calling for much thought, while others are deeply fascinating.

For example, when we glance at a clear night sky, we seek constellations of bright stars. These constellations, as striking as they may appear to us, are accidental. With time, the stars will move into new patterns. The arrangements are transient, as are the patterns of light reflected from the surface of a lake. Their detailed shapes require no explanation, any more than the pattern of sparkles on a sunlit like.

The application of probability to the imitation of physical events—we might call it "casino physics"—can provide a key. By carefully constructing a model and watching it on a computer, we can demonstrate the emergence of form out of chaos, and shape from what had been erratic. These are the goals of later sections of this book.

To take another example from my field, astronomy, even the casual observer will see that certain regions of the sky have far more than their equal share of stars. We say the stars are "clustered" on the sky. In the 18th century, John Michell, an English clergyman, showed that a cluster such as the Pleiades could not be the result of mere accident, and the concluded that its stars must be physically connected. We now have abundant evidence that he was right, and that the stars were formed as a group. The details of the internal pattern will shift as the stars move about, but the group as a whole will persist much longer than a typical constellation. Orion and the Big Dipper, for example, will move out of shape long before the Pleiades vanishes as a cluster.

This is one example of using the ideas of probability and chance in deciding what needs to be explained. Its application to star clusters is described in more detail in Chapter 11.

P.2 SOURCES OF PROBABILITY STATEMENTS

The process of which we perceive the world plays an important role in determining what we see and what we seek to explain. We may recognize several types of organizing principles at work in our perception, such as the following.

(a) Imposition of Geometric Forms

On a clear night, we have the impression of forms outlined by the brighter stars, such as the constellation we call "The Big Dipper" and the rectangular shape of the main body of Orion. These impressions are based on a repertoire of geometrical shapes that we carry in our "mind's eye." In our attempt to organize the events of our experience and to reduce them to a manageable set of perceptions, we project these shapes onto the sky and we declare a "match" when we find agreement, although the stars know nothing of squares. The selection is subjective. If we are surprised to see a square pattern among the stars, as in the constellation of Pegasus, this surprise is probably rooted in a naive—and incorrect—sense that randomness ought never result in patterns that match simple geometric forms.

The imposition of geometrical forms is useful to the scientist. In describing the motions of the planets, we often refer to Kepler's idea that they move in *ellipses*, although modern measurements and calculations show that their motions do not take them along precisely elliptical orbits. Such an idealization, based on the abstract geometrical notion of an ellipse, has been extremely useful over the centuries. On the one hand, it has provided a succinct *description* of planetary motion; with a small set of numbers we can describe the ellipse and the motion of a planet to a high accuracy. Another advantage of the ellipse is that it provides a *target for theoretical work*. When Newton was able to show that his law of gravitation would predict elliptical orbits, he knew he was on the right track, even though the details of the actual motions were much more complex. The abstract model is not the physical reality, but the behavior of the physical system imitates trajectories constructed from the model. And the model has *predictive power*. With it we can compute the approximate positions of the planets on any future or past date.

(b) Division of the Physical World into Elementary Units and Collections

Many objects are composed of large numbers of similar elements. A star cluster is a physical example of the mathematical concept of a *set*, which is loosely defined as a collection of objects with some property in common. (These concepts are described mathematically in Appendix 2.) The images of the stars on a photograph or raindrops on a pond also comprise two sets. We may speak of such sets as collections of elementary objects—stars or raindrops. Each of the elementary objects can, in its turn, be decomposed into a set, as the raindrop can be decomposed into atoms, and so on down to elementary particles.

(c) Discrete Events

Just as we divide the material world into atoms, we divide the world of our experience into *events*. An event is, for example, the flip of a coin, Roughly speaking it is an atom of experience. Collections of events, such as a series of coin flips, are the object of probability theory.

(d) Sample Space and Probability

With an event, we associate an *outcome*, and all possible outcomes make up the *sample space*. For the coin flip, the sample space is the pair of outcomes: *heads* and *tails*. No other outcomes are possible. As another example, consider the division of the bright stars between the northern and southern halves of the sky. We check off the stars one at a time, and our sample space, again, has two parts: each star is in the south or it is in the north. Appendix 1 gives the formal description of sample space.

(e) Relative Frequency and Probability

When we have finished checking the stars, we find a certain number of them, say N and S, in the two regions of sample space. If we divide each number by the total number $N + S$, we can derive their relative frequencies. Designate with the letter X the fraction of stars in the northern half of the sky, $X = N/(N + S)$. Then $1 - X$ is the fraction in the southern half. These fractions can be used to establish a variety of probability statements. For example, "If I pick a star at random on the sky, it will have have the probability X of being in the northern half."

This is the method by which the physicists and mathematicians introduce the idea of probability, and the formalism is described in Appendix 1.

P.3 PITFALLS OF INTUITIVE PROBABILITY STATEMENTS

Our intuitive notions may lead us astray when we do not follow the rules of probability carefully. For example, if we attempt to discuss single events before having defined the sample space and the precise meaning of the associated probability statements, trouble is in store. Here are some examples.

(a) Undefined Reference Sample

When we classify the events of our experience, we may implicitly use categories (divisions of sample space) that are not what they seem to a casual glance. If these categories do not happen to match the nature of the probability involved, we can be misled. We may suppose we are using one sample space when in fact we are using another.

As an example, consider the set of numbers 173492 and contrast it with the set 555555. The first looks "random" and the second does not. The reader will assume that I had a particular pattern in mind when I listed this string of 5s. The existence of such patterns can be a source of trouble when we come to evaluate the "probability" of each of these numbers.

In a certain sense, both numbers are equally probable, and our offhand statement that one is more probable than the other is incorrect. We cannot poperly say that one number is more probable than another. We must focus on the *process* by which the numbers were generated. Suppose we have constructed a noisy machine that can produce six-digit numbers essentially "at random." If we were to keep track of the output of such a machine, and we let it operate until it had generated billions of numbers, we would find the first (173492) just as often as the second (555555). Each would occur roughly one-millionth of the time. The apparent "coincidence" required to generate the number 555555 is no less likely than the event required to generate any other specific six-digit number.

Our intuitive response went wrong when we said 555555 was less likely, because we did not judge the likelihood of 555555 simply as a set of six digits. Rather we implicitly set up two categories of six-digit numbers, those whose digits are identical, $NNNNNN$, and those whose digits are not identical. Clearly there are only ten members of the first set and a million members of the second. On the basis of these numbers, the likelihood that an arbitrary six-digit number will belong to the smaller set is ten in a million. This is the basis of our intuitive response that 555555 is an unlikely number.

The discrepancy arose from the fact that we established the categories after we had seen the numbers, not before. That is, we did not consider the pattern $NNNNNN$ until after we had seen the number. This is called an "after-the-fact" (or *a posteriori*) probability calculation. The result of such a calculation is misleading because we have *established the criteria* for probability by exploiting knowledge of the number we are to consider. To give another example, if we consider the number 123246, our intuitive sense of whether this was a randomly generated number will depend on whether or not we recognize that the last three digits are the first three digits multiplied by two. If we *did* recognize this pattern, we are likely to conclude that the number was not randomly selected. But such patterns can, with a little diligence, be found in any string of digits.

If we permit ourselves enough freedom in defining what is likely and what is not, we can make any event appear nearly impossible. The aim of science is to focus on events that require explanation, so the distinction between likely and unlikely is crucial. Let us look again at the number 173492. If we look long enough, we will find something about the number that we can call surprising, and then we can say, "What a coincidence!" That is, we can find a pattern that it matches in some way. Such a pattern need be nothing more elaborate than the fact that the difference of the first two digits equals the number of integers while the sum of the second pair of digits equals the second digit, and the difference of the last pair equals the second digit, and so on. We must not be allowed to take this *a posteriori* pattern as evidence that the number is improbable.

Probability statements that depend on after-the-fact recognition of patterns will mislead us. This is one of the most common mistakes in dealing with probability estimates.

Perhaps I am sensitive to this type of error because of a teenage experience of my own. It will serve to illustrate the point. Several of us were shooting arrows, and the owner of the equipment announced that the other two of us would have a contest. A conventional target was mounted on a garden seat that had a wooden trellis behind it. I took my turn and achieved a mediocre score. Another boy shot his arrows and missed the target on all his trials. I naturally assumed he would be the loser, but when we retrieved his arrows, we found that one of them had pierced one of the wooden slats in the trellis and had split it. This event was so unexpected that the boy who had set up the game declared the other boy to be the winner. No amount of objecting on my part would convince him that the events was purely accidental and was not worth a point. "The arrow had to go somewhere," I insisted. "But, why there, in that strange spot?", was the response. "Why not?," would have been an appropriate reply, but by this time we were not listening to each other. (I suppose that my words would have been more persuasive if the equipment had been mine!)

A counterexample will provide the final emphasis. The film, "Robin Hood," which I saw as a boy, has an archery contest in which one archer sends his arrow to the center of the bull's-eye—which was his obvious goal, implicitly agreed upon before he actually shot the arrow. He is then followed by Robin Hood, who miraculously splits the previous arrow with his own and makes the audience ecstatic. Robin Hood had also implicitly announced his goal before shooting, not afterward, so his feat was truly worth a reward.

(b) The Law of Averages

Another pitfall is the misapplication of the "law of averages." If we flip an honest coin and it comes up "heads" three times in a row, how many of us would not be tempted to bet on "tails" on the next throw? If we persist in betting that the future trials will tend to restore the average, we will gradually discover that the concept of average applies only to the set of past events. It says nothing useful about a particular future event, such as the next throw.

A similar confusion can arise in the study of eruptive variable stars, such as novae and supernovae. Statistical studies of the Milky Way and other galaxies outside the Milky Way have shown that, on the average, supernovae occur at a rate of two per century in most spiral galaxies. This figure is derived from a study of thousands of galaxies, and it is a statistical result that probably applies to our own. Now the most recent supernova recognized in our galaxy was Tycho's, which occurred nearly 400 years ago. This does not mean we are more likely to see a supernova this year than we were 350 years ago. If the events are random, they are equally likely every year. In the same way, a fair coin that has been coming up "heads" is still just as likely to come up heads next time.

P.4 THE EMERGENCE OF FORM

In looking for explanations of form in physical systems, we find two principal types. One is essentially *static*; the other is *dynamic*.

Examples of *static form* are a snowflake and an automobile tire, whose shapes are maintained by electrical forces among the constituent molecules. Other examples are the round shape of the Earth and the internal layering of a star, such as the Sun. These shapes are molded by the internal forces of gravity and pressure among the particles making up the bodies of these objects.

On the other hand, a dynamic form is evanescent. A puff of photons emitted from the center of a star, for example, will produce an increase of light in a predictable curve when the photons have wandered to the surface. In much the same way, the horses in a race tend to spread out on the track in the course of the race.

Many such forms are only vaguely understood, and we cannot attempt to explain them thoroughly in this book. But our introduction to the behavior of random events will provide glimpses into the emergence of such forms

REFERENCES

Monod 1972
Stanley and Ostrowsky 1986
Stevens 1974

PART ONE

SIMULATIONS IN DISCRETE SPACES

1

HEADS AND TAILS

This chapter introduces computer simulations of a random process and describes a method of generating and testing a table of random-seeming numbers.

1.1 CLOSING IN ON RANDOMNESS

The result of a random process is shown in Table 1.1, which displays the outcome of flipping a real coin (Kerrich 1946). Successive outcomes are listed, as in a page of text, from left to right and top to bottom. Although this table was generated by a random process, it is not completely uniform. The pattern of numbers shows an irregular structure and occasional "runs" in which one type of outcome is repeated a large number of times. "Random" does not mean uniform on a small scale. It implies uniformity of structure on a large scale and a lack of regular structure on the small scale.

If we divide a randomly generated table into quarters, we should find nearly equal numbers of "heads" in each. If we cut the table into sections and reassemble them in an arbitrary fashion, the general appearance of the table will not be affected. The same would be true if we were to read the table backward. In fact, there is nothing about the table that can tell us which was the start and which was the end of the coin flips.

Exercise 1.1
Before reading ahead, carry out the following brief exercise. It will give you some insight into the nature of a random process and the law of averages.

On a sheet of lined paper mark a 20 by 10 array of 200 squares. In successive squares

TABLE 1.1   2000 Successive Flips of an Ordinary Coin (0 = Tail, 1 = Head)

```
00011101001111101000110101111000100111001000001110
00101010100100001001100010000111010100010000101101
01110100001101001010000011111011111100110110010101 1
01010000011000111001111101101010110100110110110 1 10
01111100001101100010100100000101001111110110101 011
10001100011000110001100110100100001000011101111000
11111100000000011010110100111110111100100101010 1100
11101101110010000010001100101100111110100111100010
00010011010111010101100111110110010000011010111 11
11010001111110010111111001110011111111010000100000
00001111100101010101111000011101110010001101000011 11
11000101001111111101101110110110111011010010110110011
01010011011111110010111000011110111111110000001001001
01001110111011011011111110000010101010101010101001001
11101101110011100000001001101010011001000100001100
10111100010011010110110111001101001010100000010000
00001011001101011011111100010110010100001110011 0011
11100101101100001100010011000100010001100100000 1001
01000011100000011101101110011000111001101010100 1011
01000001110110100001000111001001110000101000000000 10
10010001011000010010100011111101101111010101010000
01100010100000100000000010000001100100011011110 1010
11011000110111010110011001011110001011011010101 10 110
0000101101110101010000011100111000110100011101 1101
10001101110000010011110001110100000101000011111 10100
00111111111111101010100100110001011110010101010001 111
11000110101010011010010101111000011101111011001 1001
11111010000011101010111101101011100001000101101001
10011010000010111110101110101111000001011 0010
00110110101111101011100010100110110011000011000011000
01010011000110100111010000011001100011101011100001
11010111101111010110110111100111101110001101101 0000
0101111010011101100100111000111101100001111001 1111
011010111011100110111100001001111001111010100100 10
101000110101011011000011111000001100000000100111 0101
10001011101000101111110111100000111111101100000010
1011111101110001000011000011000111110100111011 0000
00001111011100011101010001011000110111010001110 111
100000100001101000001010000101010001011000101111 00
00101110010111010010110010110100011000001110000111
```

Source: Reproduced from J. E. Kerrich, *An Experimental Introduction to the Theory of Probability*, Copenhagen: E. Munksgaard (1946).

(from left to right and top to bottom, as though writing a page of text) make up a random sequence of the letters H and T, imitating the outcomes of flipping a coin. Try to do it quickly, without stopping to ponder the meaning of the word "random."

A fully random sequence shows no signs of memory. That is, the coin's behavior on each flip is not influenced by the outcome of any previous flip. Such a sequence is called a "Markov chain." The successive flips are independent of each other. This, of course, assumes that we pick up the coin and give it a good toss, so it tumbles a few times. In the process of these tumbles, the coin behaves momentarily in a way that is unpredictable. It does not disobey the laws of physics; it merely moves in a way that is too complicated to predict. The outcome of the flip is so sensitive to the details of the way we start the flip—how high it goes, how hard the table is, and factors of that kind—that we would need to make impossibly precise measurements in order to have adequate data for the computer to carry out a trajectory analysis. So the randomness in a series of coin tosses arises during each flip, when the coin goes momentarily out of hand.

Exercise 1.2
List three quantitative methods of testing your mind-generated table and the coin-generated table for the properties of randomness. For example, we should expect roughly equal numbers of heads and tails.

The question arises, if a process is random, how can we possibly say anything useful about it? To answer this question, we must focus on a series of outcomes, not just a single trial. In fact, we cannot say anything definite about a single event except to list the possible outcomes. In this case, our list would be *outcomes = (heads, tails)*. The list in parentheses is the "sample space" of the process. When we collect the results of a large number of trials we will find certain features, or patterns, in the frequencies of occurrence of portions of the sample space. These patterns are statistical and they depend on the nature of the experiment and not on the details of the particular sequence of trials.

As an example of such a pattern, Figure 1.1 shows another way of displaying the results of a series of flips of a "fair" coin—one that is equally likely to show *head* or *tail*. Distance along the horizontal axis measures the number of tosses that have occurred, and the vertical axis measures the fraction of those tosses that have come up heads; we call this fraction p_H, the relative frequency of heads. (A program to perform this simulation is described in the next section.)

The relative frequency fluctuates widely at first and then tends to the horizontal line $p_H = 0.5$, as N increases. The relative amplitude of the fluctuations appears to decrease steadily, and this suggests that we can find an *expectation value*, $\langle p_H \rangle = 0.5$, toward which the relative frequency converges as the number of trials increases without limit. We will use the brackets, $\langle \rangle$, to indicate these expectation values.

We will assume that the quantity $\langle p_H \rangle$ is a statistical property of the random series of tosses that will recur if we carry out another very long set of tosses. This contention is not strictly provable. In fact, each time such a finite experiment is carried out we obtain a slightly different result. Only a hypothetical infinite sequence of trials will strictly lead to $\langle p_H \rangle$.

Having found one statistical property of this experiment, we might look for others. For example, we might count the number of times that a head occurs between two tails, $\cdots$ THT $\cdots$, and the number of times a tail occurs between two heads, $\cdots$ HTH $\cdots$. We call these "runs of length one." If we anticipate some of the rules of probability to be discussed later and proceed by intuition, we can predict the frequency of such runs from the assumption that the coin is

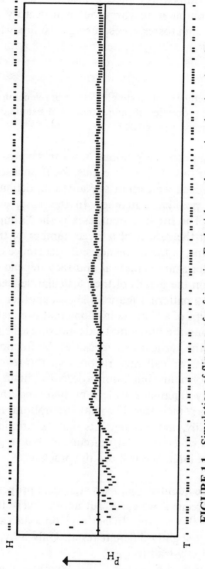

FIGURE 1.1 Simulation of flipping of a fair coin. Each tick on the top is a head and each tick below is a tail. The running fraction of heads is indicated by the height of the ticks in the middle. The expected mean, 0.5, is indicated by the horizontal line.

unbiased: $\langle p_H \rangle = 1/2$ and $\langle p_T \rangle = 1 - \langle p_H \rangle = 1/2$. Each of these sets of three consecutive outcomes has a probability $p = (1/2)(1/2)(1/2) = 1/8$, so the probability of an isolated head or tail is twice this, or $1/4$. This is the expected frequency of runs of length 1, and we write this result as

$$p(1) = 1/4.$$

Likewise, the expected frequency of runs of length two will be one-half as great, or

$$p(2) = 1/8,$$

and by induction, the probability of a run of length L for a fair coin is

$$p(L) = (1/2)^{L+1}. \tag{1.1}$$

Exercise 1.3
(a) Count the runs of different lengths in your mind-generated series and compare the results with the theoretical expectation using a histogram like that in Figure 1.2.
(b) Rearrange your series by reading it in columns instead of rows and compute a new histogram of runs lengths. Why might you expect it to differ from the original?

Although we cannot reliably predict individual outcomes, we can predict statistical properties of series events. This is the nature of probability theory. In the following, we drop the use of brackets, $\langle \rangle$, around the empirically derived expectation values and simply speak of "the probability of an event."

Exercise 1.4
Suppose we carry out a string of N tosses. We have seen that if the string is long ($N \gg 1$), the probability of a run of length $p(L) = (1/2)^{L+1}$. What is it when N is not large? With

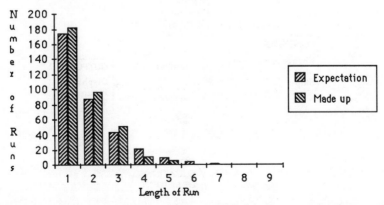

FIGURE 1.2 Theoretical and the average distribution generated by seven students, each writing a series of 100 heads and tails. There is a tendency for the students to make the heads and tails alternate, so the mind-generated frequency of short runs is unrealistically high and that of long runs is low.

$N = 3$, for example, the formula must fail because no run can exceed $L = N = 3$. In this case, there are eight possible sequences.

(a) List them and compute the frequencies of runs of different lengths.

(b) Divide Table 1.1 into sequences of $L = 3$ tosses, prepare a table of run-lengths, and compare it with your theoretical prediction.

1.2 DEVELOPING A PROGRAM TO SIMULATE COIN FLIPS

In this section, we describe the *Pascal* programming of a prototype computer simulation. In later chapters we refer to the results of similar simulations without going into detail.

The design starts from a schematic description of the physical system, a listing of the variables and parameters, and a statement of the equations that transform the system from time-step i to time-step $i + 1$. (This transformation can be considered as a series of mappings over a sequence of small time intervals.)

When the formal description is finished, we ask what sorts of controls the user should be given over the operation of the system and what sorts of data input are required to define the state of the system at the start of the simulation. Then we ask what graphical and numerical outputs are desired.

In all of this, we try to look ahead and anticipate possible changes, because the development of a computer program is always a series of trials and improvements. The program is shown schematically in Figure 1.3.

The heart of any simulation is the computation that maps the system from one time-step to the next. This is indicated by EQUATIONS OF MOTION in the diagram. For a dynamical simulation, the mapping is based on a

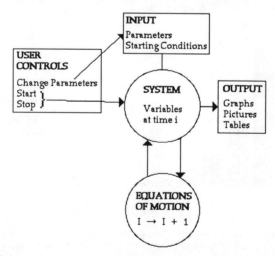

FIGURE 1.3 Schematic representation of computer simulation indicating the various elements that must be constructed. See text for discussion.

finite-difference formulation of equations which describe the time-changes of the variables. For a coin flip, it is impractical and unnecessary to solve the complete equations describing the twisting and turning of the coin from the instant the finger starts moving to the time of the last bounce on a flat surface. In this, and many similar cases, it is sufficient to use a "stochastic" description of the process. We adopt a particularly simple form and assume that the individual flips form a Markov series with the following properties:

(1) Each outcome is independent of the previous tosses.
(2) The sample space of possible outcomes is defined as a list of mutually exclusive states: *Outcome* = (*head, tail*).
(3) A probability can be assigned to each portion of the sample space. Each probability is a positive fraction between 0 and 1 and the sum of the probabilities is 1.
(4) The probability assigned to *head* is p_H, and as the outcome of a coin flip must be either *head* or *tail*, we can write $p_T + p_H = 1$, or $p_T = 1 - p_H$. Thus for an assumed value of p_H, we can evaluate p_T.

Simulating the flip means choosing randomly among the possible regions in the list that enumerates the sample space. Figure 1.4 shows a sample space for a fair coin, represented as an area divided into halves, marked *heads* and *tails*. Our task is to choose one of the areas randomly for each flip. (This formalism is dicussed more fully in Appendix 2.)

Most computer libraries have a word, such as RANDOM, that will permit this selection after a little additional programming by the simulation designer. This word is called once for each coin flip, and it generates a *pseudo-random* sequence of numbers that have the statistical properties of randomness. (Such a sequence should not be used without validating its behavior, and in the next section we describe this process in detail.)

We design a procedure, FLIP, that will return head or tail in a random sequence. The details of this procedure will depend on the output of the function

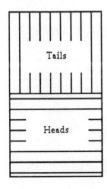

FIGURE 1.4 Schematic representation of sample space for tossing an unbiased (fair) coin. The two possible outcomes are given equal area, indicating that they have equal probability.

RANDOM. If RANDOM produces negative and positive numbers with equal probability, and we wish to simulate a fair coin, we may use the expression

If RANDOM > 0 then outcome := head

Else outcome := tail;

To handle a more general case, let us suppose that RANDOM returns numbers that are uniformly distributed in the interval LOWER $\leqslant$ RANDOM $\leqslant$ UPPER. Then we may define an auxiliary function, RANDREAL, that returns a number that is uniformly distributed in the interval $0 \leqslant$ RANDREAL $\leqslant 1$. Such a function can be defined by

RANDREAL = abs((RANDOM-LOWER)/(UPPER-LOWER));

the procedure FLIP can then be based on the conditional statement

If RANDREAL < p_H then outcome := head

else outcome := tail;

By changing the value of the parameter p_H, we can alter the bias of the simulated coin.

The results of successive trials can be stored in an array, TOSS[J], defined in terms of the enumerated list, SampleSpace:

TOSS = ARRAY[1...NumberofTosses] of SampleSpace;

This array may consist of a string of 0s and 1s or head and tail, depending on how the regions of the sample space are designated.

This array can then be analyzed for its statistical properties. For example, a two-bar histogram showing the relative numbers of heads and tails can be constructed by defining the two-entry array,

RESULTS = ARRAY[head...tail] of integer;

When the simulation is completed, the RESULTS array can be loaded by scanning the array TOSS[J] as follows:

For J = 1 to NumberOfTosses do

If TOSS[J] = head then

RESULT[head] := RESULT[head] + 1

else RESULT[tail] := RESULT[tail] + 1;

Then RESULTS [head] is the number of occurrences of a head.

A somewhat more satisfying approach is to update such histograms after every trial so the results can be plotted as the simulation proceeds. This makes the simulation more interesting to watch and gives a dynamic sense of the accumulation of the statistics.

In designing simulations, provision must be made for restarting independent runs (which requires clearing the arrays) and for plotting all relevant data. Tabular output is helpful for quantitative studies, but the qualitative features are best seen with graphics.

1.3   VALIDATING THE SIMULATION

Testing the validity of a stochastic simulation is not as direct as it would be for a deterministic simulation. Due to the random and finite nature of the computer calculations, the theoretical expectations are never met exactly. However, the use of several complementary tactics will often reveal programming errors or shortcomings of the formulation.

Before testing the overall simulation, it is necessary to prove that the random-number generator is adequate.

(a)   Testing the Random-Number Generator

The tests we describe are based on comparing statistical properties of a series with the theoretical predictions based on the assumption that the generation of the series is a truly random process. Experience shows that no single test can reveal all possible idiosyncrasies of a random-number generator. Each test is a necessary condition for acceptability, but it is not sufficient by itself to prove acceptability. So we must apply several tests of different types.

Period   All pseudo-random-number generators repeat themselves sooner or later. The period of a random-number generator may be determined by noting the first several numbers, x_1, x_2, x_3,... and then stepping through the series until this series is found again. The period should be at least twice as great as the length of a series to be used in any simulation.

Distribution   A randomly generated sequence ought to contain numbers more or less uniformly from all parts of the permitted range. A series may be tested by assigning each of its members to one of n bins. If $J = 1, 2,..., n$ are the labels of the bins, the assignment of each member of the series, $x_i (0 < x_i < 1.0)$, can be computed from $J = \text{INT}(nx_i) + 1$. On each occurrence of J, the bin count, $B(J)$, is incremented, so $B(J) \rightarrow B(J) + 1$.

Table 1.2 and Figure 1.5 show the result of generating 50 random digits (0–9) and putting them into the corresponding bins. The observed numbers in each bin cluster about the expected mean, $\langle B \rangle = N/n = 5$, and when the test was performed several times, there was no evidence that any particular bins

TABLE 1.2   Sample Frequencies of Pseudo-random Digits

Digit	0	1	2	3	4
Frequency	7	8	5	6	4

Digit	5	6	7	8	9
Frequency	5	5	2	5	3

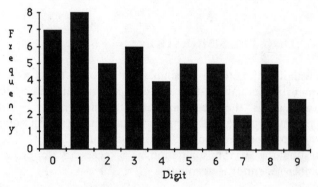

FIGURE 1.5 Frequencies of digits generated by a pseudo-random-number generator. Fifty digits (0–9) were counted, and we ask whether this distribution has the statistical features of randomness.

consistently receive more than their fair share of the counts, namely, 1/10 of the total.

Qualitatively, this map of bin populations is encouraging, but are the fluctuations from bin to bin of the right size?

Quantitative Measure of Goodness of Fit   To develop a quantitative measure of performance we may use an approximate form of the "chi-squared test," described in a private communication from James Stapleton dated October 21, 1984. This test evaluates the fluctuations, or the spread, about the mean and compares it with the theoretically expected value. (We need not concern ourselves with the theory, but the test uses the normal approximation to the chi-squared statistic.)

Define the function, W as

$$W = \frac{1}{\sqrt{2n}} \left(\sum_{J=1}^{n} \left\{ \frac{[B(J) - \langle B \rangle]^2}{\langle B \rangle} \right\} - n \right) \tag{1.2}$$

where the definition $\langle B \rangle = \sum B(J)/n$, and where the sums are taken over the bin populations. If the magnitude of W is less than unity, $|W| < 1$, we infer that the populations conform to the behavior of randomness. A large value implies that the populations appear nonrandom. For example, $|W| > 2$ will occur in only 5 percent of the cases for random tables, and $|W| > 2.58$ will occur in only 1 percent of the cases.

To understand the behavior of W qualitatively, suppose the distribution were perfectly uniform and all bin counts were precisely equal to the average. (This would occur if we pushed the random-number generator to its full period, so each number appeared exactly once.) In this case, all the terms in the curly brackets in W would be zero, and $W = -\sqrt{n/2}$ which is large and negative. The indicates that the generator is being pushed too far. On the other hand, if the bin counts deviate widely from the average, this sum becomes very large, and W becomes large and positive. This indicates that the numbers are lumpy and unevenly distributed.

Exercise 1.5
For the results in Table 1.3 with $n = 10$, show that $W = 0.98$, which is consistent with randomness.

Sequential Correlation   The tests described so far have concentrated on the statistical properties of the random numbers considered as a set, without regard to their order of occurrence. We must also be on the lookout for patterns in the order of appearance. A variety of tests may be applied to detect correlations among members of the series.

For example, the series can be used to simulate a string of coin flips, and the resulting distribution of run-lengths can be compared with the theoretical distribution.

Alternatively, the series can be used to generate a fictitious set of poker hands, and the results can be compared with the theory described in Chapter 2. Or the distribution of ascending or descending runs can be compared with theory. See Ripley (1987) for references to this test.

(b) Tactics for Testing the Simulation

After validating the random-number generator, the next task is to test the simulation as a whole. The most reliable way is to carry out a detailed hand calculation for two or more complete time-steps, examining each procedure in detail. After some experience, however, it is possible to shortcut this process and test the program output itself.

Two types of test ought to be considered. On the one hand, the simulation program can be used to predict known results under simplifying assumptions or special choices of parameters. For example, the run-length distribution can be verified for various values of p_H. Often it is possible to run the program with extreme values of the parameter and quickly validate its limiting behavior.

On the other hand, the program can be made to sidestep the random-number generator and carry out a fully deterministic calculation that may be verified independently.

Each simulation will require a different strategy, and a wise approach is to assume the program is incorrect until proved otherwise.

1.4 GENERATION OF RANDOM NUMBERS

The process by which computers produce random numbers is not truly random, because a digital computer is a deterministic machine—barring failure. So we must be satisfied with deterministic algorithms, and the series they produce are called "pseudo-random" for that reason. With a bit of care, these can be made to imitate random processes and generate numbers with most of the earmarks of randomness.

The RANDOM functions supplied with computer libraries are often adequate for typical tasks, but an appreciation of their general nature and their pitfalls is essential to the designer of simulations.

This discussion is confined to the most common type—the *linear congruential generator* (LCG)—that generates each successive term, X_{i+1}, from its predecessor. To see how such forms are constructed, we start with the simple linear exression

$$X_{i+1} = aX_i + c \qquad (a > 0, c \geqslant 0). \tag{1.3}$$

The successive results comprise an ascending series of numbers that can be represented by a series of points on the positive half of the number line, as is illustrated in Figure 1.6. In this form, the series does not appear random. We need a method of scrambling the output. One way to achieve this is to imagine the number line to be subdivided into relatively short segments of length m, where m is less than the coefficient a. (This ensures that the point jumps out of its present segment at each step.) As the point is mapped along this space, its position within each segment will be an erratic function, and this position can be considered as a pseudo-random number. Numerical values can be assigned if we suppose each segment to be scaled from 0 to m.

Such a partition of the number line into segments of length m can be achieved quite easily. We modify the equation slightly and introduce the modulus function, defined as follows:

$$X \text{ MOD } m = X - \text{INT}(X/m)m. \tag{1.4}$$

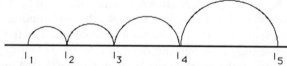

FIGURE 1.6 Mapping along the number line by multiplication. Each successive number is a fixed multiplicand of the preceding.

In this expression, $\text{INT}(X/m)$ is the integer part of the quotient, derived by cutting off the fractional part. [For example, $\text{INT}(3.1) = 3$.] $X \text{ MOD } m$ is the positive remainder when m is subtracted from X as many times as possible. For example, $11 \text{ MOD } 3 = 11 - 3 - 3 - 3 = 2$. Note that the result of MOD always lies between 0 and $m - 1$.

Rewrite the mapping as

$$X_{i+1} = (aX_i + c) \text{ MOD } m. \tag{1.5}$$

This is a general form for the LCG, and it produces a series of integers in the range $0 < X_j < m - 1$. This mapping is illustrated schematically in Figure 1.7.

An example of an LCG is the function

$$X_{i+1} = (5X_i + 3) \text{MOD } 8, \tag{1.6}$$

which generates the series

$$1, 0, 3, 2, 5, 4, 7, 6, 1, 0, 3, 2, 5, \ldots.$$

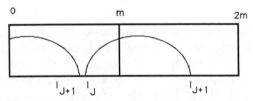

FIGURE 1.7 Mapping with the modulus function returns each point to the interval $0 < X < m$.

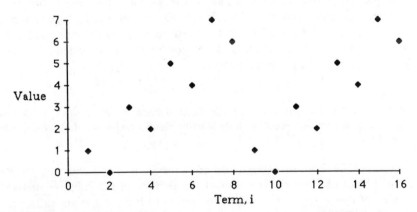

FIGURE 1.8 Successive values in the series $X_{i+1} = (5X_i + 3) \text{MOD } 8$. The odd-numbered terms increase uniformly ($x_i = i$, for i odd) and each even-numbered term is one less than the preceding odd-numbered term ($x_i = x_{i-1} - 1$ for i even). This a poor state of affairs.

TABLE 1.3   Linear Congruential Generators of the Form $X_{i+1} = (5X_i + c)\text{MOD}\,8$

Expression	X_i	Period
$5X + 1$	1, 6, 7, 4, 5, 2, 3, 0	8
$5X + 2$	1, 7, 5, 3, 1, 7, 5, 3	4
	4, 6, 0, 2, 4, 6, 0, 2	4
$5X + 3$	1, 0, 3, 2, 5, 4, 7, 6	8
$5X + 4$	1, 1, 1, 1, 1, 1, 1, 1	1
	2, 6, 2, 6, 2, 6, 2, 6	2
	3, 3, 3, 3, 3, 3, 3, 3	1
	4, 0, 4, 0, 4, 0, 4, 0	2
	5, 5, 5, 5, 5, 5, 5, 5	1
	7, 7, 7, 7, 7, 7, 7, 7	1
$5X + 5$	1, 2, 7, 0, 5, 6, 3, 4	8
$5X + 6$	1, 3, 5, 7, 1, 3, 5, 7	4
	2, 0, 6, 4, 2, 0, 6, 4	4
$5X + 7$	1, 4, 3, 6, 5, 0, 7, 2	8

This series (Figure 1.8) starts off with a haphazard appearance, but it repeats itself every eight terms. It has a "period" of 8 because the modulus of the series is 8. There cannot be more than eight integers in the series and the series is deterministic so each appearance of a particular integer must be followed by a uniquely determined integer. Hence the series must repeat itself with a period no longer than its modulus. The series also has a shorter period, as can be seen from the figure.

The properties of a LCG depend sensitively on the parameters (a, c, m). A series that generates all the m distinct integers $(0 < n < m - 1)$ during each period is said to have a "full period." Table 1.3 illustrates some of the series with $a = 5$, $m = 8$, for various values of the additive constant c. If we extend the series beyond its period, the pattern repeats.

Exercise 1.6
Verify with a hand calculation that the LCG defined by $X_{i+1} = (5X_i + 4)\text{MOD}\,8$ produces different series depending on the starting value X_1. (The starting value is called the "seed.")

Several of the series in Table 1.3 do not have a full period and they generate only a subset of integers. The sum of the periods of the subsets equals the modulus of the series. For example, the series with $c = 4$ has six subseries of period 1 or 2. The seed determines which subseries is generated. Clearly, we want a LCG with a full period.

It can be shown (see Ripley 1987) that a LCG will have a full period if and only if its coefficients obey all of the following conditions:

Conditions for Full Period

(1) c and m have no common divisors greater than unity.

(2) a MOD $p = 1$ for each prime factor p of m.

(3) a MOD $4 = 1$ if 4 divides m.

To verify these rules for the examples of LCGs in Table 1.3, we note that $a = 5$, $m = 8$, and $p = 2$. Therefore the first rule is obeyed only when c is odd. Furthermore, 5 MOD $2 = 1$, so the second rule is obeyed for all the examples. Finally, 5 MOD $4 = 1$ and 4 does divide 8, so the third rule is obeyed. Taken together these rules imply that full periods will occur for all odd values of c, as we easily verify for small c.

Exercise 1.7

Guided by the rules above, construct a LCG with a full period of 7.

We can alter the simple determinism of the series if we add together two or more distinct series, as described by Wichmann and Hill (1987). As a simple illustration, consider two LCGs, the first with a period of 9,

$$X_{i+1} = (7X_i + 7)\,\text{MOD}\,9, \tag{1.7}$$

which (with seed $= 1$) generates the series $5, 6, 4, 8, 0, 7, 2, 3, 1$, and the second with a period of 8,

$$Y_{i+1} = (5Y_i + 5)\,\text{MOD}\,8, \tag{1.8}$$

which (with seed $= 1$) generates $Y = 2, 7, 0, 5, 6, 3, 4, 1$.

The integers in each of the series, X_i and Y_i, are called up with equal frequency, so they are uniformly distributed over the intervals 0–8 and 0–7, respectively. This means that, if a member of one of these series is selected at random, it will have equal likelihood of lying anywhere in the corresponding interval. In seeking a way to combine the series we might be tempted to merely add the

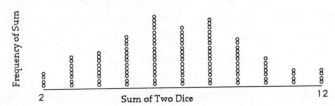

FIGURE 1.9 The distribution of the sum of two random variables is not the same as the distribution of the individual variables. In this example, a pair of six-sided dice was thrown (in simulation) 100 times and the frequencies of their sums were tabulated. Individually, the dice show uniform distribution in the range 1–6, but their sums give the triangular distribution in the range 2–12 shown in this figure.

series together to form $Z_i = X_i + Y_i$. But if X_i and Y_i are uniformly distributed, their sum will have a triangular distribution. Intermediate values will be more frequent, as illustrated by the sums of two dice in Figure 1.9 and verified in Exercise 1.8.

Exercise 1.8
The terms defined by the sum of the two LCGs given above, $Z_i = X_i + Y_i$, will lie in the interval 0–15. Construct a computer program to evaluate successive terms, and construct a histogram of the frequency of occurrences of each value, Z_i, showing that these sums are more likely to lie near the middle of its range.

In order to avoid this nonuniform distribution, Wichmann and Hill devised the following procedure:

(1) Divide each series by its corresponding $m - 1$, obtaining numbers between 0 and 1; for example, $x_i = X_i/8$, $y_i = Y_i/7$.

(2) Add the series together, term by term, obtaining the new series, $Z_i = x_i + y_i$, which will lie in the range $0 < Z_i < 2$.

(3) Take only the fractional part, $z_i = Z_i \text{MOD} 1$. This will renormalize the numbers to a uniform distribution in the range $0 < z_i < 1$. (The effect of taking the fractional part is to add two sloping distributions to form a single flat one.)

By tailoring three LCGs to the word-length of the computer, it is possible to create additive random-number generators with much longer periods. Wichmann and Hill combined three LCGs with $c = 0$. They selected the following:

$$X_{i+1} = \text{MOD}(171 X_i, 30{,}269),$$
$$Y_{i+1} = \text{MOD}(172 Y_i, 30{,}309), \qquad (1.9)$$
$$Z_{i+1} = \text{MOD}(170 Z_i, 30{,}323).$$

They then followed the three-step prescription above, normalizing each by dividing by its modulus and summing the results, obtaining the intermediate variable T_{i+1},

$$T_i = (X_i/30{,}269 + Y_i/30{,}309 + Z_i/30{,}323). \qquad (1.10)$$

The random variable R_i was then obtained from the fractional part of T_i as follows:

$$R_i = T_i - \text{TRUNC}(T_i). \qquad (1.11)$$

Wichmann and Hill also devised a special manipulation to avoid exceeding the integer capacity of the 16-bit machine, and their formulas for evaluating the individual LCGs are shown in the spreadsheet listing in Appendix 3.

The period of the series R_i is very great—approximately 7×10^{12} terms before repeating, and at the rate of one million numbers per second, a computer would require several months to cover a single period of the sequence.

The Wichmann–Hill algorithm represents a highly satisfactory pseudo-random-number generator for use with 16-bit computers. According to the authors, this generator passed its tests an appropriate fraction of the time. To have passed the tests all the time would, of course, have been too good to be true and would not have been characteristic of a truly reliable random-number generator!

Exercise 1.9
Program the Wichmann–Hill random-number generator and apply two of the tests described in this chapter.

REFERENCES

Bennett 1976
Kalos and Whitlock 1986
Kerrich 1946
Knuth 1981
Ripley 1987
Rubinstein 1981
Wichmann and Hill 1987

2

GAMES IN DISCRETE SPACES

The calculation of probabilities often involves counting the points in an elaborate sample space, and there are numerous devices for carrying out this process. We describe some of them in this chapter, starting with the game of craps and concluding with poker. (See Appendix 1 for definitions of unfamiliar terms. Readers familiar with combinations and permutations may simply scan this chapter for a short review.)

2.1 SUMS OF TWO DICE

Suppose we throw a pair of ordinary six-sided dice and tabulate the sums. Figure 1.9, in the previous chapter, shows the outcome of a computer simulation of 100 throws (trials, in the language of probability), and the frequency is highest for the intermediate values. To compute the theoretical expectation, we proceed in two steps:

(1) Tabulate all possible outcomes for each trial. (This is called mapping "the sample space" for the experiment.) As each die has six possible points (1–6) there are $6 \times 6 = 36$ possible outcomes for each toss of two dice. A portion of them are listed below, along with the corresponding sums.

Die 1	1	1	1	1	1	1	2	2	2	2	2	2	….
Die 2	1	2	3	4	5	6	1	2	3	4	5	6	….
Sum		2	3	4	5	6	7	3	4	5	6	7	8	….

TABLE 2.1 Sums of Two Dice

Sums	2	3	4	5	6	7	8	9	10	11	12
Relative frequency	1	2	3	4	5	6	5	4	3	2	1

(2) Count the number that correspond to each sum, and construct Table 2.1 listing the sums and the relative frequencies for each sum. For example, there is only one way for the sum to be 2, namely, when both dice show 1. We indicate this portion of the sample space as $\{1, 1\}$. On the other hand, there are two ways to achieve 3—when the dice are $\{1, 2\}$ and $\{2, 1\}$—corresponding to the fact that two regions of the sample space give the same sum.

Exercise 2.1
Write out the complete sample space for tossing two cubic dice and verify the entries in Table 2.1.

Exercise 2.2
Compute the expected frequencies for the sums of two four-sided (tetrahedral) dice.

The expected frequencies define a triangular shape, as might have been inferred from the simulation in Figure 1.9. Seven is expected to be the most common sum.

2.2 THE GAME OF CRAPS

The following description of craps is taken from Weaver (1982) although the game has many variants:

> A player rolls two dice. If the sum of the two is 7 or 11, the roller wins at once; if the sum is a 2, 3, and 12, the roller loses at once. If the sum is 4, 5, 6, 8, 9 or 10, he rolls again until he either "makes his point" and wins, by repeating the initial number, or "craps out" and loses, by rolling a 7.

Figure 2.1 is a diagram of the game, indicating the three possible outcomes for the first throw of the dice: win, lose, or establish the "point" for the continuation. We first ask for the probability of winning on the first throw of the dice, and these may be calculated from Table 2.1.

To compute the expected relative frequency (or probability) of winning on the first throw, we note that this will require a 7 or an 11, and we indicate these events by the symbols S7 and S11, meaning the sum is 7 and the sum is 11. Thus a win occurs in the event (S7 $\cup$ S11). (The symbol $\cup$ is the *union* and this can be read "in the event that the sum is 7 or 11." See Appendix 1 for a discussion of this notation.) A loss occurs in the event (S2 $\cup$ S3 $\cup$ S12), and we

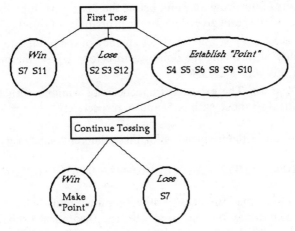

FIGURE 2.1   Schematic representation of craps game described in text.

have for the expected frequencies of these events

$$p(win\ on\ first\ throw) = p(S7) + p(S11) = 2/9,$$
$$p(lose\ on\ first\ throw) = p(S2) + p(S3) + p(S12) = 1/9.$$

These two events are disjoint (mutually exclusive), so the expected frequency of games that are won or lost on the first throw is the sum of their individual frequencies, or 1/3. And as

$$p(game\ ended\ on\ first\ throw) + p(game\ not\ ended\ on\ first\ throw) = 1,$$

it follows that the expected frequency of games not ending on the first throw is $1 - 1/3 = 2/3$.

If the game is neither won nor lost on the first throw, the point is 4, 5, 6, 8, 9, or 10, and the game is continued until the player either craps out or makes the point. We now examine the expected frequency of winning in each of these cases, using the numbers in Table 2.1.

Point = 4: We have $p = 3/36 = 1/12$ for the frequency of this point. The tosser wins if another 4 is thrown before a 7. There are three ways for the win and six for the loss, and the other events are of no consequence. Thus the relative frequency for a win at this stage is 3/9 or 1/3. Multiplying this by the relative frequency of having the point = 4, gives 1/36 for the probability of winning this way.

Point = 5: We have $p = 4/36 = 1/9$ for the frequency of this point. There are four ways to throw the point again and six ways to throw the 7. Thus the probability of throwing the point before the 7 is 2/5. Multiplying this by the frequency of the point gives 2/45 for the probability of winning this way.

Point = 6: We have $p = 5/36$ for the frequency of the point and 5/11 for the probability of throwing the point again before a 7, so the probability of winning this way is 25/396.

Point = 8, 9, or 10: The expected frequencies of winning with these points are identical to those with 6, 5, and 4, respectively.

The summation of the frequencies for these different ways of winning gives

$$p(win) = 0.2222 + 2(0.0278 + 0.0444 + 0.0631) = 0.4929.$$

Figure 2.2 shows the outcomes of 400 computer-simulated games, and the fractional outcomes may be compared with the theoretical probabilities. The craps game is usually played at even odds, and we see that the person throwing the dice (who is playing against the house) has a slightly unfavourable situation. In some casinos, it is permitted to bet against the tosser, but in this case the rules are slightly altered to prevent the house being taken to the cleaners.

The flowchart for a computer program to simulate a craps game is shown in Figure 2.3. The first steps are to specify the number of sides on the dice and the number of games to be played. This is done by keyboard input. Then the arrays that will be used to store the outcomes are all set to zero and the screen

Results of 400 games:
Overall fraction of wins =0.488
Fraction of naturals=0.230
Fraction of craps =0.133

Point	Occurrences	Wins	Fractional wins
4	32	12	0.375
5	39	16	0.410
6	50	24	0.480
8	63	28	0.444
9	40	14	0.350
10	31	9	0.290

FIGURE 2.2 Results of playing 400 simulated games of craps. Upper portion shows graph of outcomes and lower portion tabulates the fractional wins with various points.

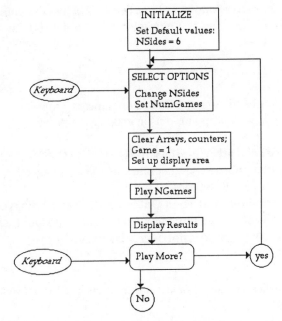

FIGURE 2.3   Flow chart for simulation of craps game.

is prepared for display of the game. The program then plays the required number of games, displays the results, and requests the user to indicate with the keyboard whether more games are to be played.

From the programmer's standpoint, the most interesting part of this program is the part that plays the series of games. The schematic code is the following:

```
While Game < NGames +1  {if game counter is less than limit}
      DoFirstThrow  {player craps out or wins or establishes "point"}
{toss until player makes point or loses}
      If not gameIsLost then PlayRestofGame
      Display results
      Game = Game + 1  {increment game counter}
End While  {End of game loop}
```

The subroutine **DoFirstThrow** can be written as follows:

```
NPoint = 0  {Initialize value of Point}
DiceTotal=INT(NSides*RND+1)+INT(NSides*RND+1)  {Use random}
{number to generate sum of two integers in the range 1 to NSides}
If DiceTotal=7 or DiceTotal=11 Then Outcome=1: Return  {Player wins}
```

If DiceTotal=2 or DiceTotal=3 or DiceTotal=12 Then Outcome= -1: Return
{Player loses and game ends. Otherwise, Point is set and game continues}
NPoint = DiceTotal
Outcome=0
Return

The continuation of the game, **PlayRestofGame**, is programmed as a loop of tosses which ends with crapping out or making the point:

> **100 DiceTotal=INT(NSides*RND+1)+INT(NSides*RND+1)**
> {Use random number to generate sum of two integers}
> {in the range 1 to NSides}
> **If DiceTotal=Npoint Then Outcome = 1: Return** {Player wins}
> **If DiceTotal=7 Then Outcome = -1: Return** {Player loses}
> **GOTO 100** {If game is not over, make another toss}

Exercise 2.3
Construct a computer program to simulate the craps game and use it to test the theoretical predictions.

Exercise 2.4
Imagine a "craps" game played with a pair of tetrahedral dice and test the following rules. A player wins on the first throw with a S5 and loses with S2 or S8. If S3, S4, S6, or S7 are thrown, the play continues until the player "craps out" with a S5 or wins by throwing the point again. Compute the odds for winning this game, and show that the probability of winning is 0.494, so this game is slightly "fairer" than a conventional game with six-sided dice described above.

2.3 CALCULATING THE SIZES OF SETS IN LARGE SAMPLE SPACES

In dealing with sample spaces that are too complicated to be counted explicitly we require formulas that do the counting for us. Two types of counting technique are often useful.

(a) Permutations

Imagine that $N = 15$ numbered balls have been put into a bag. We draw five of them one at a time and jot down their order. They are put back into the bag and five are drawn again—and again. How many possible sequences of five markers are there? (Or equivalently, how many different five-letter combinations can be formed from a set of N letters if we are not permitted to use any letter twice?)

The answer will be indicated with the symbol $\prod(N, n)$, representing the "permutations" of N objects taken n at a time. In our case, we take $N = 15$

objects, $n = 5$ at a time. Each sequence is a permutation, as it considers the particular order in which the markers were drawn from the bag. Let us count them. There are N ways of choosing the first member, $N - 1$ ways of choosing the second, $N - 2$ ways of choosing the third, and so on. The total product is $N(N - 1)(N - 2)(N - 3)(N - 4)$ and this is the answer.

By induction we have at once the general expression for the number of permutations of N objects taken n at a time:

$$\prod(N, n) = \frac{N!}{(N - n)!}. \tag{2.1}$$

Exercise 2.5
   (a) How many arrangements are there for six birds sitting on a bench (Figure 2.4)?
   (b) How many pairs can be selected from among six people?

(b) Combinations

A *combination* is a collection in which sequence plays no role. In this regard, combinations differ from permutation. For example, there are three combinations of three stars taken in pairs: $\{1, 2\}$, $\{1, 3\}$, and $\{2, 3\}$. On the other hand, there would be six permutations of three stars taken in pairs because $\{2, 3\}$ and $\{3, 2\}$ represent different permutations—in which the order of appearance is significant. We use the symbol $C(N, n)$ to designate the number of combinations that may be created from N objects when taken in sets of n. Therefore $C(3, 2) = 3$.

The total number of permutations of N objects taken n at a time, $\prod(N, n)$, equals the number of combinations, $C(N, n)$, times the number of permutations in each combination. This suggests a way to compute $C(N, n)$. Since there are $n!$ permutations in each group of n, we have

$$\prod(N, n) = n! C(N, n). \tag{2.2}$$

With our previous result for the number of permutations, this gives at once

$$C(N, n) = \frac{N!}{n!(N - n)!}. \tag{2.3}$$

FIGURE 2.4 How many arrangements are there for six birds on a fence? How many pairs can be formed?

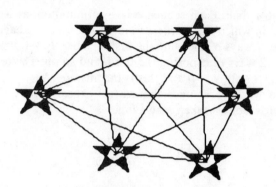

FIGURE 2.5 Six stars in an imaginary sky. How many lines may be drawn between pairs?

Note the symmetry, $C(N, n) = C(N, N - n)$, which says that the number of combinations of ten objects taken three at a time equals the number of combinations when taken $10 - 3 = 7$ at a time.

As an example, consider the array of six stars shown in Figure 2.5. They represent bright stars in an imaginary sky, and we are to construct constellations from lines joining them. The first question is, how many lines may be drawn? The answer is

$$C(6, 2) = \frac{6!}{4!2!} = 15. \tag{2.4}$$

Now how many triangles may be formed? We cannot simply take all sets of three lines, because many of the lines do not intersect at a star. Rather, we compute how many combinations of three stars may be formed. We find $C(6, 3) = 20$.

Exercise 2.6

Show that 63 groups consisting of one to six stars may be concocted from a sky of six stars. How many nonoverlapping constellations can you form?

2.4   THE POKER TEST FOR RANDOM NUMBERS

Simulating the dealing of poker hands can be a useful, if somewhat tedious and elaborate, method of testing random-number generators. In any case, an understanding of the probabilities of various hands can be useful. Table 2.2 lists the nine possible types of poker hand and the number of ways of being dealt each hand. After explaining a few of these entries, we leave the remainder to the reader.

TABLE 2.2   Poker Hands

Hand	Number of Ways
Straight Flush	40
Four of kind	624
Full house	3,744
Flush	5,108
Straight	10,200
Three of a kind	54,912
Two pair	123,552
One pair	1,098,240
No Pair	1,302,540
Total number	2,598,960

(1) *Total Number of Possible Hands*: Fifty-two cards taken five at a time gives $C(52, 5) = 2,598,960$ combinations. (At the rate of dealing five hands per minute, it would take about 1 year to deal this number of hands.)

(2) *40 Straight Flushes*: In each suit, there are ten cards that can be the lowest in a five-card straight: Ace–10. Hence $4 \times 10 = 40$.

(3) *624 Four-of-a-Kinds*: There are 13 possible quadruples. For each of these, the fifth card in the hand must be filled with one of the remaining $52 - 4 = 48$ possibilities. Hence $13 \times 48 = 624$.

(4) *3744 Full Houses*: This hand consists of three cards of one "spot" and two cards of another spot. There are 13×12 ways to pick the spots. Next, we note that the triad can be filled in four ways, because any one of one four suits can be excluded. The pair may be filled in $C(4, 2) = 6$ ways. Hence there are $13 \times 12 \times 4 \times 6 = 3744$ full houses. And so it goes....

We can use these numbers to test the randomness of pseudo-random-number generators. One procedure would be to deal hands of five cards with the random-number generator, and this can be achieved by generating a series of numbers from 1 to 13 and partitioning it into groups of five. Then count the fractional number of different types of hand. A much simplified version is obtained if we merely count the fractions of hands with two, three, or four of a kind. The results should then be compared with the predictions using a modification of the chi-squared test discussed in Chapter 1.

REFERENCES

Knuth 1981
Weaver 1982

3

THE RANDOM WALK IN ONE DIMENSION

This chapter introduces the most important process in statistical physics, the random walk. We start with the simplest case, motion along a line, such as a bead sliding randomly on a wire. Figure 3.1 represents two samples of random walks along the y-axis, with time increasing to the right. The walker moves randomly up or down, as though governed by the outcome of a coin flip. This type of behavior is a model for "Brownian motion," the irregular, jiggling path of atoms in a gas. It also has applications to the motion of photons and of impurities in a crystal.

We ask: How far do we except the walker to move in 10 steps?

3.1 DIFFUSION OF MOLECULES IN A PIPE

A narrow pipe will serve to confine our walk to one spatial dimension. Because molecules are invisible, we imagine a frog at the center of a 500-cm section of pipe that is open at both ends. The frog moves exactly 20 cm each time it jumps, so it could escape from the pipe in 25 jumps if it moved in a single direction (see Figure 3.2). But it doesn't. Instead, it jumps in randomly selected directions. Occasionally, it will take several jumps toward the nearer end of the pipe, but then it reverses temporarily. We ask how many jumps we ought to expect the frog to take before it emerges from the pipe.

This is a simple form of the "random-walk" problem, and in this chapter, we consider a walk in a one-dimensional space with probability p of stepping to the right and $q = 1 - p$ of stepping to the left.

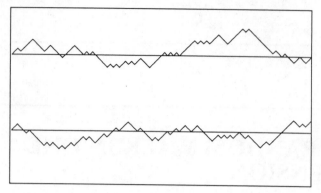

FIGURE 3.1 Diagram of two random walks in which time increases to the right. Each step is randomly chosen up or down and all steps are of the same length. Such diagrams are simplified representations of the motions of molecules in a gas.

FIGURE 3.2 A frog starts at the center of a pipe and jumps randomly to right and left. Our task is to predict the average location of the frog after N steps.

3.2 THE GALTON BOARD

Figure 3.3 is a schematic diagram of a device that demonstrates such a random walk, namely, a "Galton board." This device is useful because it helps visualize the process and provides a model that can be analyzed in some detail. Balls are introduced at the top, and they tumble toward the bottom, bouncing to right or left off a pin at each level on the way. The ball always moves downward one step at each bounce, so the only random element is the motion sideways. Hence this corresponds to a random walk in one dimension. (In effect, we have turned Figure 3.1 on its side.) We consider a symmetric walk, with $p = q = 1/2$, so the ball has equal probability of moving to right or left at each step of its path.

The trajectory is a series of binary choices and the pattern of outcomes is revealed by the number of balls that have fallen into each slot at the bottom of the board. This is the pattern we wish to analyze. This particular diagram is the analogue of the net displacement of a frog after four jumps.

The position of the walker can be indicated in two ways. We can label its position by n, the total number of steps to the right, or we can use m, the net

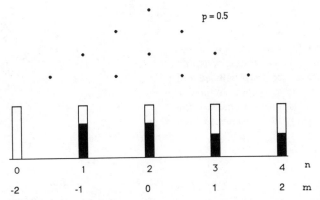

FIGURE 3.3 Diagram of Galton board. Balls bounce downward among nails into cups at the bottom. Each bounce is a binary choice (*right* or *left*). In this case, 10 balls were run through the board and the number in each cup is indicated by the height of the bar. The integers n and m indicate alternative ways of denoting the endpoints, as discussed in the text.

displacement to the right. To see the relation between $n(0 < n < 4)$ and $m(-2 < m < +2)$, note that the sum of steps to the right and to the left is N.

Exercise 3.1
Show that, in a walk of N steps, the total steps to the right, n, and the net displacement of the right, m, are related by $m = 2n - N$, or $n = (m + N)/2$.

Exercise 3.2
Carry out a simulation of the frog walk using a coin to decide whether the frog moves to right or left. Assume all steps to be of equal length, and start the frog at a distance $L = 2$ from the ends of the pipe. Stop each walk when the frog reaches the end and tabulate the walk-length L required for escape from the pipe. Repeat for a longer pipe, in which the frog is initially at a distance $L = 4$ from the ends. Repeat the simulation at least five times and determine how N increases with L.

3.3   THE PASCAL TRIANGLE EMERGES FROM THE RANDOM WALK

The short walk in Figure 3.4 can be considered as representing a small Galton board. The ball must move downward each time, as this corresponds to the advance of "time." The central position corresponds to $m = 0$. According to our assumption, $p = 1/2$, $q = 1/2$, the walker has equal probabilities of turning right or left at each branch; thus each particular path to the bottom has exactly the same total probability, $(1/2)^N$. Thus if we want to compute the likelihood of ending at a particular slot, we must count the number of paths to that slot.

FIGURE 3.4 A short walk on the Galton board. The dark circles represent nails that deflect the descending balls, and the open circles are four balls placed according to the expected endpoints. Their number is proportional to the number of equally likely routes to each endpoint.

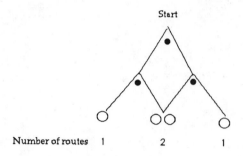

Start

Number of routes 1   2   1

FIGURE 3.5 Pascal triangle. Each entry is the sum of the numbers lying adjacent in the line above. The horizontal rows correspond to various N and the diagonals correspond to various n.

```
                                     n
                                     0
           N   2N              1     1
           1   2           1    1    2
           2   4       1    2    1    3
           3   8    1   3    3    1    4
           4  16  1  4    6    4    1   5
           5  32 1  5   10   10    5   1
```

When we count the number of routes to each point from the apex, we will have constructed one horizontal row of the Pascal triangle. A portion of the triangle is shown in Figure 3.5.

We label each row by the latter N, representing the number of steps required to reach that row. Position in each row is labeled by n, the number of steps to the right. Each entry in the table is labeled $C(N,n)$, and the algorithm for constructing the Pascal triangle is that each entry is the sum of the two upper adjacent entries, or

$$C(N,n) = C(N-1, n-1) + C(N-1, n). \tag{3.1}$$

This relationship is also the rule for computing the number of paths to each point at a lower level in the Galton board, and this correspondence explains why the Pascal triangle can be used as a map of the number of paths to each point in the Galton board. We find the full expression for $C(N,n)$ in Chapter 4.

3.4 THE PATTERNS OF A RANDOM WALK

In order to study the pattern more thoroughly, we need more data, so it is time to construct another computer program.

Exercise 3.3
Construct a computer program along the lines described below.

The program will carry out a series of **NW** walks, each of length **LW**. The walker's location will be called **Place**, and each walk will start from **Place** = **0** and will move one step to the right or left at each time-step: **Place** = **Place** + **1** or **Place** = **Place** − **1**, depending on a selected random number. (In this respect, the program is identical to the coin-flipping program of Chapter 2.) At the end of each walk, after **NL** steps, the value of **Place** will be used to increment an array that keeps track of the endpoints: **Nend(Place)** = **Nend(Place)** + **1**. This array will then provide a histogram of endpoints.

The program will have three sections:

(1) *Initialization*

Choose **LW**, the length (number of steps) of each walk.

Choose **NW**, the number of walks.

Define the array **Nend(j)** to keep track of the endpoints.

The value of j covers the range $-\text{LW} < j < \text{LW}$. For computer languages that do not permit negative array indices, we add LW and use a displaced index $j' = j + \text{LW}$. Clear the array by setting **Nend(j)** = **0** for all j.

(2) *Carry Out the Walks*: This is performed with a pair of nested do-loops. The outer loop, $i = 1$ to NW, keeps track of the successive walks, and the inner loop, $j = 1$ to LW, does the series of steps for each walk. Schematically, the program for the series of walks is

```
For i = 1 to NW do
        Place=0  {Put walker at origin}
        For j=1 to LW do  {Move walker depending on random number}
            if RND > p then Place = Place + 1 else Place = Place - 1
        Next j
        Nend(Place) =Nend(Place) + 1   {Update array after each walk}
Next i
```

(3) *Display the Results*: Each value of NEND(j) represents the number of walks ending at Place = j. These can be plotted as a histogram. A simple way to do this is to print a string of asterisks, ∗∗∗, corresponding to the integer in each array cell. For more elaborate plots, such as Figure 3.6, we need to determine the scale of the histogram. This can be done by scanning the array to find the largest entry, MAXFREQ:

```
Maxfreq = 0
For  j = -lw to lw
        if nend(j) > Maxfreq  Then  Maxfreq = Nend(j)
Next j
```

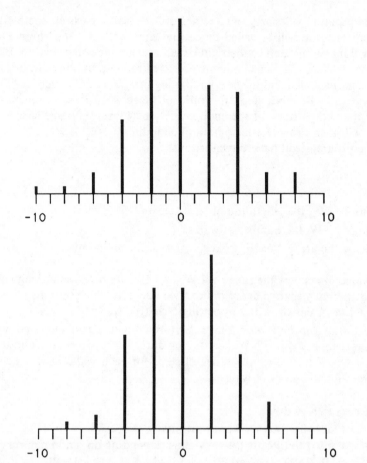

FIGURE 3.6   Two simulations of 100 walks of length 10 steps. The histograms show the distributions of endpoints.

Several examples are shown in Figure 3.6 for LW = 10 steps. What can we predict about the shape of the histograms of endpoints? First, we expect the mean displacement to be zero, because we expect as many steps in one direction as in the other. The detailed shape of the distribution can be derived from the Pascal triangle. The entries on the Pascal triangle represent the number of paths to a given endpoint, and the sum over all endpoints increases as 2^N. For our case, $p = 1/2$, the relative numbers in the Pascal triangle give the shape of the distribution of outcomes. We merely need to divide each row by the normalizing factor, 2^N, to get the shape for a walk of length N.

Exercise 3.4

Carry out your own simulation and construct histograms for symmetric walks ($p = 1/2$) of various lengths. For each length, LW, how large must NW be before the pattern becomes clear?

Exercise 3.5

(a) We refer to the histograms in Figure 3.6 as the "observed" numbers of walks ending at each point and use the symbol w to represent these numbers. Now extend the Pascal triangle to a walk of 10 steps and compute the expected numbers, $\langle w \rangle$.

(b) Apply the modified chi-squared test, described in Chapter 1, to the differences. The outcome of this test indicates whether the simulations imitate a truly random walk. Evaluate the sum

$$X = \frac{1}{\sqrt{2n}} \left(\sum_{j=1}^{n} \left\{ \frac{(w - \langle w \rangle)^2}{\langle w \rangle} \right\} - n \right)$$

and refer back to Chapter 1 for the significance of X.

3.5 EXPECTED SPREAD OF THE ENDPOINTS

Imagine putting several frogs into pipes of different lengths. How can we crudely predict the effect of the length of the pipe on the duration of the walk? (Refer back to your results in Exercise 3.2.)

The endpoints of a large number of walks of length N form a cloud along the x-axis, and the cloud grows progressively broader as N increases. We may suppose that this cloud represents the endpoints of the random hopping of the frog, and when the edge of the cloud reaches to the ends of the pipe, the frog will have a good chance of escaping. (There is inevitably a certain amount of vagueness in defining what we mean by the "edge" of the statistical cloud. We can sharpen this definition later.)

Figure 3.7 shows one example of how the size of this cloud increases as more jumps are made. In this figure, the tracks of many random walks starting from the same point and moving with constant step-length to right or left are plotted, with time increasing downward. A careful examination of many such walks shows the remarkable fact that the average width of the cloud increases according to a simple law:

The mean square of the width D increases linearly with the number of steps N; that is, $\langle D^2 \rangle \sim N$.

This simple result (which we derive more rigorously later) is one of the most useful and universal relations in statistics. We may rewrite it by defining the root-mean-square width of the pattern of endpoints, D_{rms}, as the positive root,

$$D_{rms} \equiv \sqrt{\langle D^2 \rangle}.$$

It increases as the square root of N,

$$D_{rms} \sim \sqrt{N}. \tag{3.2}$$

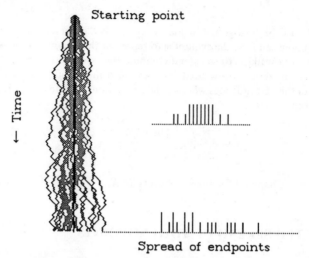

Starting point

Spread of endpoints

FIGURE 3.7   Twenty random walks of length 220 steps starting at a single point as shown with time increasing downward. The upper and lower histograms show the locations of walkers at the midpoint and the finalpoint. The increasing spread of walkers is obvious. This spread increases as the square root of the number of steps, as discussed in the text.

The root-mean-square distance, D_{rms}, is a measure of the mean displacement without regard to sign. When this distance is comparable to the initial distance from the end of the pipe, the frog is likely to have exited the pipe.

In order to understand the expansion of the cloud of points and derive this expression more rigorously, let us look at a series of steps. The frog starts from the origin and moves back and forth along a line in steps whose lengths are a constant, $|L|$, but whose signs are random, for example, $X_1 = -L$, $X_2 = L$, $X_3 = L$, and so on, with equal probability. The situation after a particular walk of N steps can be represented by the series

$$S_N = X_1 + X_2 + \cdots + X_N.$$

We ask: What will be the statistical nature of the sum after a large number of such walks of length N?

One feature is clear at once. Because there is no distinction between right and left, there is no tendency for the ensemble of endpoints to deviate in one direction or the other. Thus the average position of the endpoints must be at the starting point.

Formally, we see this as follows. Suppose a large number of walks are performed. In each walk, the values of $X_1, X_2, \ldots$ are selected randomly. The algebraic signs of each step are randomly distributed, so the average of the each step is zero:

$$\langle X_1 \rangle = \langle X_2 \rangle = \cdots = 0.$$

Hence the *average* net displacement vanishes:

$$\langle S_N \rangle = \langle X_1 + X_2 + \cdots \rangle = 0. \tag{3.3}$$

But what about the *spread* of the individual endpoints? As a measure of this spread we take the mean square of the deviations from the average net displacement. This is called the *variance* and we evaluate it as follows, taking a very short walk, $N = 3$, as our first example. Here are six randomly generated walks, each listed vertically in a column:

Walk Number	1	2	3	4	5	6
Step						
X_1	1	1	1	-1	-1	-1
X_2	1	-1	1	1	-1	-1
X_3	-1	-1	-1	1	-1	1
Sum S_3	1	-1	1	1	-3	-1

Averaging the values of S_3, we find $\langle S_3 \rangle = 1 - 1 + 1 + 1 - 3 - 1 = -2/6 = 0.33$, so the right and left steps did not quite balance in this instance. Summing the squares of S_3, we find

$$\sum S_3^2 = 1 + 1 + 1 + 1 + 9 + 1 = 14,$$

so the average is

$$\langle S_3^2 \rangle = 14/6 = 2.333.$$

In the general case we see that, after a large number of walks of N steps, the expected square displacement is

$$\langle D^2 \rangle = \langle S_N^2 \rangle = \langle (X_1 + X_2 + \cdots)(X_1 + X_2 + \cdots) \rangle$$
$$= \langle (X_1^2 + X_1 X_2 + \cdots + X_2 X_1 + X_2^2 + \cdots) \rangle.$$

Only the squared terms remain on the right-hand side; all of the others, such as $\langle X_1 X_2 \rangle$, vanish in the average, because X_1 and X_2 are not correlated, and for any selected value of X_1, there are as many positive as negative values of X_2. We have

$$\langle D^2 \rangle = \langle (X_1^2 + X_2^2 + \cdots) \rangle. \tag{3.4}$$

There are N terms, and they are all equal to L^2, so we have for the expected square of the net distance the following relation, which is one of the most important in statistical physics:

$$\langle D^2 \rangle = NL^2. \tag{3.5}$$

The mean square of the net displacement equals the number of steps times the square of the step-length.

Thus the expected net displacement after 100 steps is 10 times the step-length. After 1000 steps, it is 31 times the step-length.

We are now in a position to discuss the fate of the frog (which we sent in to represent a molecule) introduced at the start of this chapter. We suppose that it will escape, on the average, when $\langle D^2 \rangle = NL^2$, if D is the initial distance from the end of the pipe and L is the jump-length. So we can solve this expression for the expected number of steps required for escape. We find

$$N = \left(\frac{D}{L}\right)^2. \tag{3.6}$$

This number increases as the *square* of the distance from the end of the pipe, D. Thus long pipes require disproportionately more steps.

We can also estimate the *total distance* the molecule must be expected to move before escaping. This distance is given by the expected number of steps multiplied by the length of each step, or NL. But we just found $NL = D^2/L$, so the total distance may be written

$$\text{Total distance} = \left(\frac{D}{L}\right)D. \tag{3.7}$$

This is beautiful result. Expressed in words, the frog must travel a total distance equal to (D/L) times the straight-line distance to the end of the pipe.

We also see that the fraction $D/L = \sqrt{N}$ is the number of steps that would have been required to escape if all steps were in the same direction. Thus the expected number for a random walk is just the square of the number for a straight-line walk.

Exercise 3.6
If a molecule moves across a room (10 meters) in a random walk with steps of 10^{-6} meters, how many steps would be required on average, and what is the total distance the molecule would be expected to travel?

REFERENCES

Mosteller et al. 1970
Reif 1965

4

PINBALL AND STOCHASTIC HORSE RACES

This chapter extends the results of the last chapter to random walks involving large numbers of trials and unequal probabilities of jumping to right and left. The reader is invited to start with the following exercise.

Exercise 4.1
Carry out a Monte Carlo simulation of a Galton board in which a ball drops past 7 two-way intersections into 8 bins. The probability of bouncing to the right is p, and to the left is $1 - p$. Tabulate the outcomes for sufficient runs to make the statistical pattern clear. Construct a histogram of the results in which the populations of various bins are represented by vertical bars. Do this for $p = 0.40, 0.50, 0.60$.

4.1 ANALYZING THE GALTON BOARD

The random walk on a line may be considered as a series of trials. For each trial, there are only two possible outcomes: "success" (step to the right) with probability p, and "failure" (step to the left) with probability $q = 1 - p$.

$$\leftarrow (1 - p) \qquad p \rightarrow$$

These are "binary" trials, and the probabilities are assumed constant through the series. In many situations it is reasonable to assume the successive trials to be independent of each other, so the outcome of one trial does not influence latter trials. When this is true, the specific *order* of the successes and failures will not affect the final position, only the relative *numbers* of successes and failures will count.

Suppose we consider a "success" in each trial to have a relatively small probability, for example, $p = 0.25$. A sequence with a large portion of failures will be more likely in this case than one with a large portion of successes. The sequence of four successes, {ssss}, would have a probability $(0.25)^4$, while the sequence of four failures, {ffff}, would have the probability $(0.75)^4$.

Exercise 4.2
Suppose a thumbtack is dropped repeatedly on a table and it is known to have a probability $p = 1/3$ of coming to rest with its point up. Call this event $\{U\}$ and its complement $\{D\}$. Calculate the probabilities of the following runs:

>   (1) {UUD},   (2) {UUUD},   (3) {UUUUD},
>
>   (4) {DDU},   (5) {DDDU},   (6) {DDDDU},
>
>   (7) {DUD},   (8) {DDUD},   (9) {DDDUD}.

As is demonstrated by the previous exercise, the probability of a particular sequence is independent of the order of successes and failures.

Now consider a walk of N steps, in which the walker moves n steps to the right and $N - n$ to the left. Such a walk might be represented by the string of N symbols: RRLLLRLRLLL$\cdots$ in which there are n symbols of type R and $(N - n)$ symbols of type L. The trials are assumed independent of each other, so their probabilities may be multiplied together to find the probability of the entire series. The probability of such a walk is $p^n(1 - p)^{N-n}$. The order of appearance of the Rs and Ls is irrelevant; only their total number counts.

In fact all possible arrangements of the R and L symbols are equivalent, as long as we keep n and $N - n$ constant. The number of equivalent walks (with specified n and $N - n$) can be expressed as the number of combinations of N outcomes when n of them are of one type and $N - n$ are of the other. This is given by

$$C(n, N) = \frac{N!}{n!(N - n)!}. \tag{4.1}$$

This is the number of equivalent sequences and multiplying this number by the probability of each individual sequence, $p^n(1 - p)^{N-n}$, we come to the following general result:

> **The probability of n successes in N trials when the probability**
> **of success in a single trial is p is given by**

$$B(N, n) = \frac{N!}{n!(N - n)!} p^n(1 - p)^{N-n}. \tag{4.2}$$

This is the "Bernoulli distribution"—often called the "binomial distribution." It is the most important formula in this book. Memorize it or learn to derive it.

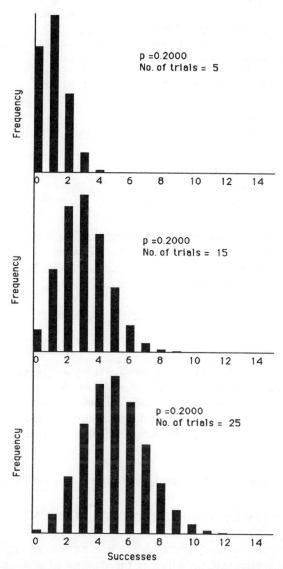

FIGURE 4.1 Bernoulli distributions for $p = 0.2$, probability of success in a single trial. Each histogam shows the relative frequencies of various numbers of successes to be expected in N trials. From the top down, $N = 5, 15, 25$. This figure shows, for example, that in a run of 15 trials, we ought to expect 3 successes to be the most likely number. Also, the frequency of 4 successes falls between the frequency of 2 successes and 3 successes.

Exercise 4.3

Compare your histograms generated in Exercise 4.1 with the theoretically expected outcome by computing the probabilities of getting various numbers of heads in 7 tosses $[P(7, n)$ for $n = 0, \ldots, 7]$. (Remember $0! = 1$.)

Hint: To do this, you will need to compute factorials, and the following BASIC statements can be used as a subroutine:

```
If N = 0 Or N = 1 Then Nfact = 1: Return
Nfact = N
Nt = N
While Nt>2
       Nfact = Nfact*(Nt-1)
       Nt=Nt-1
Wend
Return
```

The numbers that appear in the Pascal triangle (cf. Chapter 3) and the solution to the random walk, $C(N, n)$, are known as the "binomial coefficients," because they also appear in the expression for the expansion of a binomial raised to the power N. The general expansion may be expressed as,

$$(a + b)^N = C(N, N)a^N + C(N, N-1)a^{N-1}b + C(N, N-2)a^{N-2}b^2 + \cdots$$
$$+ C(N, N-n)a^{N-n}b^n \cdots + \cdots C(N, 0)b^N, \tag{4.3}$$

where the coefficients are given by equation (4.1).

Figure 4.1 shows the Bernoulli distribution of the expected number of successes with $p = 0.2$ for success in a single trial and various numbers of trials, N. Note that the highest point of the histogram moves to the right, indicating a larger number of successes, as the number of trials increases.

Exercise 4.4

By summing the bars on Figure 4.1, verify the relation: Mean number of successes $= pN$.

4.2   PROBABILITY OF A SPECIFIED DISPLACEMENT

For many problems it is handy to replace the total number of steps to the right, n, by the net displacement, m, which is the number of successes minus the number of failures. From the definition, $m \equiv n - (N - n)$, given in Chapter 3, we find

$$n = \frac{N + m}{2}. \tag{4.4}$$

Substituting in the expression $B(N, n)$ above gives $P(N, m)$, the probability of displacement m in N trials:

$$P(N,m) = \left[\frac{N!}{[(N+m)/2]![(N-m)/2]!} \right] p^{(N+m)/2}(1-p)^{(N-m)/2}.$$

This may be written

$$P(N,m) = W_{N,m} p^{(N+m)/2}(1-p)^{(N-m)/2}, \tag{4.5}$$

where $W_{N,m}$ is the number of paths of length N leading to displacement m. It is given by

$$W_{N,m} = \frac{N!}{[(N+m)/2]![(N-m)/2]!}. \tag{4.6}$$

Figure 4.2 and Table 4.1, bottom row, show the outcome of a Galton board with $N = 7$ steps onto which a total of $Q = 744$ balls were dropped. These data were obtained from a real board and are not the result of simulation.

The expected numbers, $M(m)$, are related to the total number of balls, Q, by $M(m) = QP(7,m)$.

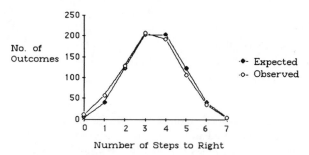

FIGURE 4.2   Observed outcomes of 744 walks on a real Galton board, compared with the expected distribution for $p = 0.5$.

TABLE 4.1   Distribution of Outcomes for a Real Galton Board ($N = 7$ Steps; $Q = 744$ Walkers)

m	-7	-5	-3	-1	1	3	5	7
n	0	1	2	3	4	5	6	7
$C(N,n)$	1	7	21	35	35	21	7	1
$M(m)$	6	41	122	203	203	122	41	6
Counted	12	58	129	206	192	107	36	4

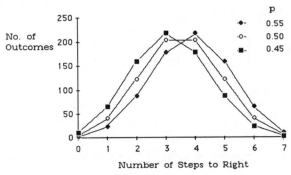

FIGURE 4.3   Expected outcomes of tilted Galton boards. Three sets of walks with various probabilities of stepping to the right, $p = 0.45$, 0.5, 0.55, for comparison with Figure 4.2.

Row 3 in Table 4.1 gives the number of routes and row 4 gives the expected number of balls, while row 5 gives the counted numbers. The computed and counted numbers do not agree, and the deviations do not appear to be random. The counted numbers are larger than the computed values on the left side ($m < 0$), and they are smaller than the computed numbers on the right side. This suggests that there was a slight preference for jumps to the left when the balls moved down the board, as thought the board were defective in some way. The simplest hypothesis is that the board was slightly tilted, since this would alter the probabilities in the two directions. Figure 4.3 shows three sets of expected outcomes computed with probabilities $(1 - p) = 0.45$, 0.5, and 0.55, respectively. (These values were computed with a spreadsheet program, as described in Appendix 3.)

Exercise 4.5
Using a spreadsheet program, determine the relative probabilities, p and $1 - p$, of a Galton board that would lead to the asymmetry shown by the graph in Figure 4.2. Adjust the value of p until you get the "best fit," as estimated by eye.

Exercise 4.6
To obtain a quantitative estimate of the goodness of fit for each of the histograms in the previous exercise, use the modified "chi-squared" criterion of Chapter 1 (and Exercise 3.5) to compare the observed and computed bin populations. Evaluate W for each value of p, and plot the value of the W as a function of p to find the best value. Judging from the value of W, does the theoretical curve provide an adequate match to the measured histogram?

Returning to our original frog, we can now ask: What is the probability, $P(N, m)$, that it will have moved a *net distance m* after N steps? In other words, what is the expected distribution of distances from its origin after a certain number of steps? This is exactly the same as the answer to the Galton board problem.

Qualitatively, we see that short walks cannot get far from the origin; therefore the density of endpoints is high in the neighborhood of the origin and it declines as the endpoints drift away.

For example, consider a point at a finite distance, say $|m| = 2$, from the origin. After a very short walk, say $N = 1$ steps, none of the endpoints can have reached $|m| = 2$, because no walks can be farther than $|m| = N$ from the origin. At $N = 2$ a small fraction of walks will arive at $|m| = 2$. For walks slightly longer than two steps, we expect a larger number, but as the walks become even longer, the endpoints spread over a wider and wider region of space, leaving fewer in the vicinity of the origin. Figure 4.4a shows the result of 100 simulated walks and the histogram on the left gives the number of times the walker was exactly two steps from the origin, either to the right or left, during the walk. The number is small at first and rises rapidly; then it falls slowly. Figure 4.4b is a similar result for $|m| = 6$, and we see that the number of arrivals at this distance from the origin rises more slowly.

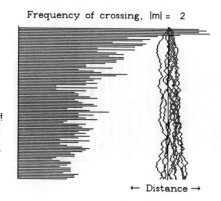

Frequency of crossing, |m| = 2

Time →

← Distance →

FIGURE 4.4a   One hundred simulated random walks along a line. The histogram on the left shows the arrivals at points 2 units from the origin. (For clarity, only ten walks are mapped on the right.) Figure 4.5a shows the expected behavior.

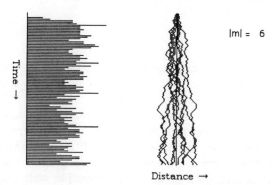

Time →

|m| = 6

Distance →

FIGURE 4.4b   Similar to Figure 4.4a, except that arrivals at points 6 units from origin are plotted. Figure 4.5b shows the expected pattern.

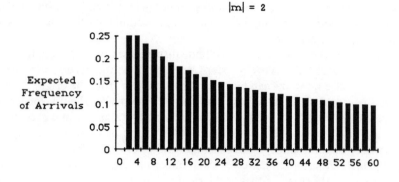

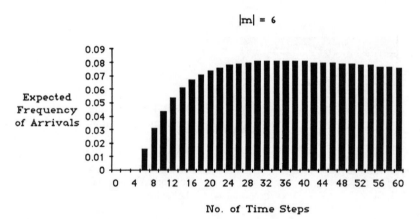

FIGURE 4.5a Expected arrivals at points 2 units from origin. Compare with simulations in Figure 4.4a.

FIGURE 4.5b Similar to Figure 4.5a for points 6 units from origin. Note the longer rise time compared to Figure 4.5a.

In order to compute the expected pattern, we use the probability of moving from one point to another in a given number of steps, $P(N, m)$, and we will take $p = 1/2$ for simplicity. We are interested in the variation of $P(N, m)$ for fixed m and varying number of steps, N, and this variation is shown in Figures 4.5a and 4.5b. The curves correspond to arrivals at two and six steps from the origin. For points near the origin, the curve immediately slopes downward; for points away from the origin, the curve remains at zero until $N = |m|$, then it rises rapidly and passes through a shallow maximum. The walk-length required to attain this maximum probability varies as N^2, as can be verified by further calculation.

4.3   THREE USEFUL SUMMATIONS

Summing the probabilities $B(N,n)$ for all possible outcomes after a walk of N steps ought to lead to unity, because this is the key property of a probability distribution. We can verify that this is the case from

$$\sum_n B(N,n) = C(N,0)(1-p)^n + C(N,1)p(1-p)^{n-1} + \cdots + C(N,N)p^n,$$

but this is just the binomial expansion, given earlier in this chapter, for $[(1-p)+p]^n$, so we have

$$\sum_n B(N,n) = [(1-p)+p]^n = 1^n = 1. \qquad \text{QED} \qquad (4.7)$$

Another useful result is the mean number of successes, $\langle n \rangle$, in N trials when the probability of success in a single trial is p. We find this with a little sleight-of-hand, starting from the general binomial expansion:

$$(p+q)^N = \sum C(N,n)p^n q^{N-n}. \qquad (4.8)$$

Differentiating with respect to p, we have for the left-hand side

$$\left(\frac{d}{dp}\right)(p+q)^n = N(p+q)^{N-1}, \qquad (4.9)$$

while the right-hand side gives

$$= \sum n C(N,n)p^{n-1}q^{N-n}.$$

Multiplying both sides by p, we find

$$pN(p+q)^{N-1} = \sum n C(N,n)p^n q^{N-n}. \qquad (4.9a)$$

Since $p+q=1$, the left-hand side is pN, while the right-hand side becomes the weighted sum of n, so we have

$$pN = \sum n B(N,n) \equiv \langle n \rangle. \qquad (4.10)$$

Thus *the mean number of successes is just p times the number of trials*. The *variance* of a distribution is a measure of its width, as described in Appendix 1, and it is defined as the mean squared difference from the mean, so

$$s^2 = \langle (n - \langle n \rangle)^2 \rangle = \langle n^2 \rangle - \langle n \rangle^2. \qquad (4.11)$$

Using the relation we just derived, $\langle n \rangle = pN$, we can evaluate the variance of the number of successes, n. We start with equation (4.9a) and differentiate both sides with respect to q. We then multiply by q, finding

$$N(N - 1)pq(p + q)^{N-2} = \sum n(N - n)C(N, n)p^n q^{N-n}.$$

Since $p + q = 1$, the left-hand side simplifies and the right-hand side may be rewritten in terms of mean values, so

$$N(N - 1)pq = \langle n \rangle N - \langle n^2 \rangle.$$

Inserting $p = 1 - q$ and collecting terms, we have

$$Npq = \langle n^2 \rangle - \langle n \rangle^2.$$

But the right-hand side is the variance, so we may rewrite this equation as

$$s^2 = Npq = Np(1 - p). \tag{4.12}$$

Exercise 4.7
Verify the derivation of equation (4.12).

An interesting property of the variance can be seen if we consider a variety of experiments in which the total number of trials, N, is constant but p takes different values. For example, the first experiment might consist of 100 tosses of short thumbtacks, for which $p = 0.2$, the second experiment might be 100 tosses of long tacks, for which $p = 0.5$, and the third might be 100 tosses with $p = 0.7$. Suppose each experiment were performed a large number of times and the variances of the outcomes were computed. Which experiment would be expected to produce the largest variance?

We find this by differentiating the expression $s^2 = N(p - p^2)$ with respect to p and setting the slope to zero, finding $1 - 2p = 0$. This gives $p = 0.5$ as the value of p for which the variance will be greatest.

Exercise 4.8
Verify this result by computing the relevant Bernoulli distributions.

4.4   PARTICLE GROWTH AS A STOCHASTIC HORSE RACE

An important application of simulation techniques is the study of the growth of particles by accretion from a dilute medium, such as a vapor. The simplest model of particle growth by accretion is to assume that each particle is repeatedly bombarded by atoms but only a small fraction stick. We can make two rather drastic simplifying assumptions: (1) ignore loss from the surface and (2) assume that each particle has an equal probability of accreting an atom in a given time interval. (This ignores the important effect of increasing area, a correction that is left to the reader.)

To help visualize this process, we can imitate it by a race among H horses. Each particle is represented by a horse, and the horses move forward by chance. Moving one step away from the starting gate is equivalent to accreting an atom. A random number from 1 to H is generated to select a horse to move ahead one step, and this process is repeated N times. Each horse is assumed to have an equal probability, $1/H$, of being selected at each step, an we are interested in the distribution of particle sizes (distances moved) after N steps. (To include the effect of area, the selection probability would be a function of the distance moved from the starting gate.)

At each step, each horse faces a binary trial: either it *will* be selected to move ahead or it *will not*. Let $p = 1/H$ and $q = 1 - 1/H$ be the probabilities for success and failure for each horse at each trial. We ask the following: In a series of N trials, what is the probability that a particular horse will achieve n successes when the probability of success in each trial is p? Phrased this way, the problem leads at once to the Bernoulli distribution:

$$B(N,n) = \frac{N!}{n!(N-n)!} p^n (1-p)^{N-n}.$$

The mean number of successes will be pN, according to the result in the preceding section, so the mean number of steps traveled by the horses is expected to be $\langle n \rangle = N/H$.

This mean value can be viewed in two ways:

(1) It is the expected mean for a large number of *races* for a given *horse*.

(2) It is the expected mean for a large number of *horses* in a given *race*.

We may summarize these by saying that $\langle n \rangle$ represents the mean outcome of a large number of independent races of length N.

Table 4.2 shows the expected outcomes for competitions among 5 horses in races of various lengths: $N = 15, 25$, and 50. The columns give the probability that a horse will make n steps in each of these races, and we note that the peak (mode) of each computed distribution lies near the mean, $\langle n \rangle$. Each entry is the probability that a given horse will end the race at point n. It can also be interpreted as the fractional number of horses that will end the race at point n. These distributions are plotted in Figure 4.6, and several features of their shapes are characteristic. First, they become broader as the length of the race is increased and the horses become more widely scattered. Second, the curves become shallower as they become broader; the area under each curve remains constant, because the number of horses is fixed. Finally, the curves are nearly symmetric about each of the mean values.

The expression for the width (square root of the variance) can be supplemented by an expression for the *relative* spread of the distribution, $s/\langle n \rangle$,

TABLE 4.2   Theoretical Solutions for Stochastic Horse Race ($H = 5$ Horses)

Number of Steps Advanced, n	Length of Race = N		
	15	25	50
0	0.035	0.0038	0.0000
1	0.132	0.024	0.0002
2	0.231	0.071	0.0011
3	0.250	0.136	0.0044
4	0.188	0.187	0.013
5	0.103	0.196	0.030
6	0.043	0.163	0.055
7	0.014	0.111	0.087
8	0.0035	0.062	0.117
9	0.0007	0.029	0.136
10		0.012	0.140
11		0.0040	0.127
12		0.0012	0.103
13			0.076
14			0.050
15			0.030
16			0.016
17			0.008

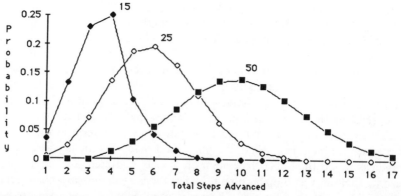

FIGURE 4.6   Expected outcomes of stochastic races of lengths 15, 25, and 50 steps. Five horses competed in each race, and the races terminated after the prescribed number of steps. The plots show the probability of advancing various numbers of steps before the race terminated. This is an example of the Bernoulli distribution.

the square root of the variance divided by the mean distance traveled. We have

$$\frac{s}{\langle n \rangle} = \frac{\sqrt{Npq}}{Np} = \sqrt{\frac{1-p}{Np}}, \tag{4.13}$$

and we see that, for constant p, the *relative* spread decreases as $1/\sqrt{N}$, the reciprocal of the square root of the length of the race. That is, the longer the race, the narrower is the *relative* spread of the outcomes. This is in distinction to the *absolute* spread, which increases with the square root of the length of the race. (Compare with the "law of large numbers" discussed in Appendix 2.)

Exercise 4.9
Simulate a number of "races" among 5 horses and compare the results with the theory illustrated in Figure 4.6.

4.5   ANOTHER PERSPECTIVE ON THE STOCHASTIC HORSE RACE

Rather than keeping track of the actual distance achieved by each horse, we can change our point of view slightly. Let us imagine that we ride alongside the race on a camera van that moves with exactly the average speed of all the horses. In this case, some of the horses will be ahead and some will be behind, and we expect to find approximately equal numbers ahead and behind us.

Let us compute the relative position of a typical horse with respect to our van. First, we express the true distance, D, of the horse along the track after N steps of length L as the sum

$$D = \sum_{i=1}^{N} Z_i. \tag{4.14}$$

In this expression, Z_i is a random variable representing the progress of the horse at each step of the race, and it takes the values

$$Z_i = L \quad \text{with probability } 1/H;$$
$$Z_i = 0 \quad \text{with probability } 1 - 1/H.$$

So D increases stochastically with jumps of length L as the race progresses.

Now to compute the progress as seen from the van that moves with the average speed, we first note that the average motion along the track is $pL = L/H$ per step. Subtracting this from the typical horse's motion, we find a new random variable for the motion,

$$X_i = Z_i - pL. \tag{4.15}$$

Using $p = 1/H$, this variable takes the values

$$X_i = L(1 - 1/H) \quad \text{with probability } 1/H;$$
$$X_i = -L/H \qquad \text{with probability } 1 - 1/H.$$

(4.16)

The new variable (representing the typical horse as seen from our moving van) describes an asymmetric random walk, in which short steps to the rear (negative X_i) occur with high probability and longer steps ahead (positive X_i) occur with lower probability. In other words, we see the typical horse gradually move behind and occasionally leap ahead.

The group of horses will spread forward and backward from the average as the race progresses, but the pattern will remain symmetric about the average—that is, the spread is the same in both directions. The shorter length, L/H, of the rearward steps, which have the greater probability $(1 - 1/H)$, exactly compensates for the greater length, $L(1 - 1/H)$ of the forward steps, which have the lower probability $1/H$. As a result, the expected spread to the left and to the right both equal the quantity $(L/H)(1 - 1/H)$.

Thus from the standpoint of the average, the process is seen as a random walk carrying the walkers in both directions.

REFERENCES

Hoel et al. 1972
Stanley and Ostrowsky 1986
Wax 1954

5

GENERALIZING THE RANDOM WALK

This chapter considers extensions of the elementary random walk to varying step-lengths and to walks in several spatial dimensions. As an example, consider the flight of a photon within a fog bank as it bounces from one droplet to another (see Figure 5.1). Photons will wander randomly until they are absorbed within a droplet or reach the edge of the fog bank and escape.

At each step along the way, the photon moves a random distance in a random direction, and we are interested in estimating the length of the photon's path and the time required for escape, and we wish to account for the varying step-length and the fact that the photon is free to move in three-dimensional space. We consider these effects one by one starting with a one-dimensional space.

5.1 UNEQUAL STEP-LENGTHS

We start with the simpler case of a walk along a single direction and we examine the effect of having different step-lengths. Suppose a walk along a line consists of steps of two magnitudes, $|A| = 1$ and $|B| = 2$. In simulating such a walk (Figure 5.2), the two step sizes would be chosen by random lot with relative frequencies $N(A)$ and $N(B)$. Once the length has been chosen, the sign of the step (to right or left) is randomly selected with equal probability for $+$ and $-$ signs.

The net displacement in a typical walk will be described by an expression like

$$D = A + B - A - B - B + A + B \cdots \tag{5.1}$$

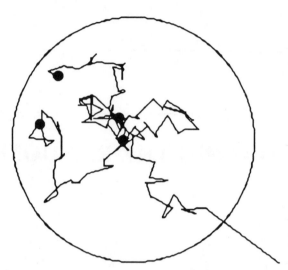

FIGURE 5.1 Random walk of photons in a circular fog bank. Steps are in randomly chosen directions and the length of each step is a randomly selected integer between 1 and 10. We wish to estimate the total length of the path and the expected distance between the origins and the endpoints, as well as the average number of steps needed to escape from the circle.

After regrouping and labeling the steps, this may be written

$$D = A_1 + A_2 + \cdots + B_1 + B_2 + \cdots \qquad (5.2)$$

where A_i and B_i randomly take the values $\{A, -A\}$ and $\{B, -B\}$, respectively. As before, we take the square of this expression and then average it over many walks. Again, most of the terms will average out, for example, $\langle A_i B_j \rangle = \langle A_i A_j \rangle = \langle B_i B_j \rangle = 0 \ (i \neq j)$.

For the remaining terms we find

$$\langle D^2 \rangle = \langle A_1^2 \rangle + \langle A_2^2 \rangle + \cdots + \langle B_1^2 \rangle + \langle B_2^2 \rangle + \cdots \qquad (5.3)$$

If we suppose there are $N(A)$ steps of length A and $N(B)$ steps of length B, the summation on the right may be written

$$\langle D^2 \rangle = N(A)A^2 + N(B)B^2. \qquad (5.4)$$

Thus we simply add together the squares of all the step-lengths, each according to the number of times it occurs. This expression can be extended to include any number of different lengths, because there is nothing in the derivation limiting it to two values.

We may define a mean step-length, S, as the length that would give the same

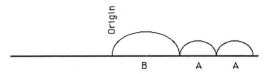

FIGURE 5.2   Random walk along a line with unequal step lengths A and B.

net displacement for the given total number of steps, $N = N(A) + N(B)$. The square of this expected value is found from

$$S^2 \equiv \frac{\langle D^2 \rangle}{N} = \frac{N(A)A^2 + N(B)B^2}{N}.$$ (5.5)

Exercise 5.1
Using the above equation, show algebraically that a random walk consisting of $N_1 = 5$ steps of length $L_1 = 2$ and $N_2 = 5$ steps of length $L_2 = 4$ would give the same expected net displacement as a random walk of 10 steps of length $S = \sqrt{10}$.

Exercise 5.2
Use the binomial expansion to show that the average squared step-length is always less than the squared average step-length—or is it?

5.2   ACCUMULATION OF ERRORS AND SUMS OF MANY DICE

The random walk of varying step-length can be applied to a wide variety of problems, and one of the most important of these is the theory of the errors of measurement. It may seem odd that there could be such a thing as a "theory of errors." Why not just eliminate the errors, if we have a theory for them? Well, in truth, we do not have a theory for the errors in detail. But we can say a great deal about the way the errors will *statistically* affect the measurements; and we can put limits on our confidence in measured numerical values.

The final error of a measured value (which is never known exactly) is the result of many smaller errors. As the errors accumulate, some cancel each other and some reinforce each other. And some are larger than others. This is just the way the steps of a random walk accumulate to a net displacement. So the final error can be visualized as the endpoint of a walk in which the directions and lengths of the individual steps are randomly selected.

Understanding how to simulate such random walks will assist in getting the most out of data when there are errors in the measurements. Here is a way to simulate this type of random walk. Suppose we toss a handful of six-sided dice, say, N of them, and count the sum of their points (Figure 5.3). (Refer back to

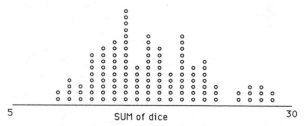

FIGURE 5.3 Sums of 5 six-sided dice. The sum is tallied after each group of five is thrown and the distribution of sums is plotted as a histogram. The expected shape is that of a bell curve.

Section 2.1 for a discussion of the expected sums of two dice.) Each die may contribute as much as 6 or as little as 1 to the sum, in much the same way that randomly distributed errors of measurement may contribute much or little to the measured value of an experimental quantity.

Exercise 5.3
Show that the expected sums are distributed according to a bell-shaped curve when n becomes large by constructing a computer program to repeat this experiment many times. Carry it out for $N = 5$, 10, and 20 dice. (See Figure 5.3 for an example with 5 six-sided dice.)

We may analyze this experiment as follows. Let D be the sum of N dice, so that

$$D = \sum_{i=1}^{N} d_i \qquad (1 < d_i < 6), \qquad (5.6)$$

where d_i is a random integer that is uniformly distributed between 1 and 6. Each value of d_i has the probability $p = 1/6$, and the mean value is $\langle d \rangle = 3.5$, as is easily verified. Now subtract the mean value and introduce the random variable $z_i = d_i - \langle d \rangle$. This new variable will be uniformly distributed in the interval $(-2.5 < z_i < 2.5)$, and it will take the values

$$z_i = \pm 2.5, \quad \text{each with } p = 1/6,$$

$$z_i = \pm 1.5, \quad \text{each with } p = 1/6,$$

$$z_i = \pm 0.5, \quad \text{each with } p = 1/6.$$

Thus the variable z_i represents a random walk with three equally probable step-lengths, $|z_i| = 2.5$, 1.5, and 0.5. Therefore the solution to our problem is that D will be distributed about the mean value $\langle D \rangle = \langle d \rangle = 3.5$ with a spread corresponding to a random walk of length N with these step-lengths. From Section 5.1, we have an expression for the rms value of the equivalent step-length,

$$S = \sqrt{\frac{0.5^2 + 1.5^2 + 2.5^2}{3}} = 1.71.$$

Exercise 5.4
Recall that the variance, σ^2, is defined as the mean squared difference, $\sigma^2 \equiv \sum (D - \langle D \rangle)^2 / N$. Using the value for S, the mean step-length given above, verify from your simulations in Exerccise 5.3 that the variance of the sums of N dice is NS^2.

A series of random walks placed end to end is equivalent to a single random walk with mixed step-lengths. The converse is also true: A random walk of N steps of length L may be broken into a series of walks of N_i steps of length L such that $N = \sum N_i$.

Exercise 5.5
From a fixed starting point, carry out a series of, say, 10 random walks of N_1 steps of length L_1 along a line (as the frog in the pipe) and evaluate the variance of the endpoints. Then, from each endpoint, carry out a series of random walks of N_2 steps of length L_2. Evaluate the variance of the new endpoints with respect to the original starting point, and verify that it obeys the relation given above for S.

5.3 TOTAL LENGTH OF A RANDOM WALK

The total distance, T, traveled during a random walk is a well-determined quantity if the step-length is fixed; it is the product of the step-length and the number of steps: $T = NL$. It is much greater than the net displacement achieved during the walk. From $NL^2 = \langle D^2 \rangle$, we have $T^2 = N \langle D^2 \rangle$, so

$$\frac{T}{\sqrt{\langle D^2 \rangle}} = \frac{\sqrt{\langle D^2 \rangle}}{L} = \sqrt{N}. \tag{5.7}$$

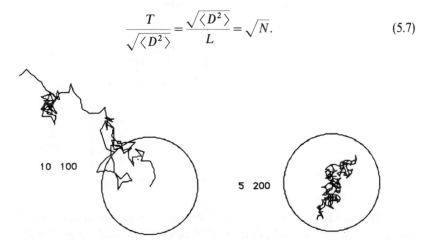

10 100

5 200

FIGURE 5.4 Two-dimensional walks of total length $T = 1000$ units. In the left-hand walk, the step size was $L = 10$ units and it was half as great in the right-hand walk.

In words, this states that the total distance, T, is to the rms net distance, $\sqrt{\langle D^2 \rangle}$, as the rms net distance is to the mean step-length, L.

Thus the rms net displacement is the harmonic mean of the mean step-length and the total distance traveled. This leads to the result that *the smaller is the step-length, the longer is the total path traveled for a given net distance.* Examples of two walks of the same total length but different step sizes are shown in Figure 5.4.

5.4   RANDOM WALK WITH A CONSTANT PROBABILITY OF STOPPING: APPLICATION TO LIGHT IN FOG

An important problem in meteorology and astrophysics is the length of the random-walk path traveled by a photon that has a definite probability of being absorbed during each step. Suppose, at each step of length L, the photon has a small probability epsilon, ε, of being stopped and a probability $1 - \varepsilon$ of undergoing another step in a random direction to right or left. We ask: What is the mean distance it will travel before being stopped?

On the average it will walk $N = 1/\varepsilon$ steps before being absorbed. The mean square displacement for an average walk of N steps will be $\langle D^2 \rangle = L^2 N$, so the root-mean-square (rms) displacement is $\sqrt{\langle D^2 \rangle} = L/\sqrt{\varepsilon}$.

Figure 5.5a shows the endpoints of 500 walks whose lengths were 2 steps, and this is to be compared with Figure 5.5b, which shows 500 walks whose lengths were statistically distributed with probability $\varepsilon = 1/2$ of stopping at each step. Note that, although the mean spread is the same in both cases, the shapes are quite different. This difference arises from the fact that the walks in the first diagram are limited to 2 steps, while the walks in the second diagram have a finite probability of extending beyond 2 steps.

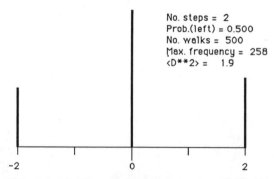

No. steps = 2
Prob.(left) = 0.500
No. walks = 500
Max. frequency = 258
⟨D**2⟩ =   1.9

-2   0   2

FIGURE 5.5a   Endpoints of 500 walks of length 2 steps. The number of walks that ended at the origin is about equal to the total number that ended 2 steps from the origin. Contrast with Figure 5.5b, in which the walk-lengths were statistically distributed about the same mean.

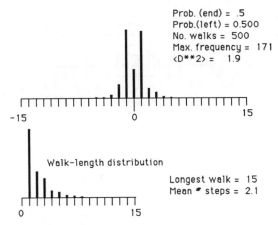

Prob. (end) = .5
Prob.(left) = 0.500
No. walks = 500
Max. frequency = 171
<D**2> =   1.9

-15   0   15

Walk-length distribution

Longest walk = 15
Mean # steps = 2.1

0   15

FIGURE 5.5b Endpoints of 500 walks in which the probability of stopping after each step is $p = 0.5$. In contrast with Figure 5.5a, the origin is not the most likely endpoint.

5.5 RANDOM WALK WITH MEMORY

Thus far, we have considered walks in which the vector displacement, x_{i+1}, at time i can be derived from the displacement at the single preceding time, i, by an equation of the form $x_{i+1} = x_i + e_i$, where e_i is a randomly selected vector. For example, in a one-dimensional walk of constant step-length, e_i is selected from the set $\{1, -1\}$. A variety of important applications arise from generalizing this relation to include the displacements at earlier times, as in the expression

$$x_{i+1} = ax_i + bx_{i-1} + cx_{i-2} + e_i, \tag{5.8}$$

where a, b, and c are adjustable parameters. The behavior of this walk will depend critically on the magnitudes of the parameters.

Exercise 5.6
Develop a spreadsheet program (see Appendix 3) to explore the behavior of random walks that obey expressions of this form. You will find a remarkable variety of behaviors. Such walks have "memory" and they can show quasi-periodic behavior, because each step depends on a series of previous steps, rather than merely on the preceding position.

5.6 RANDOM WALK ON A LATTICE

(a) Rectangular Lattice

Instead of permitting atoms to wander in any arbitrary direction, suppose we permit displacements only to adjacent points on a two-dimensional rectangular grid, as illustrated in Figure 5.6 and 5.7. This imitates the wandering of impurity atoms in a crystalline or metallic lattice.

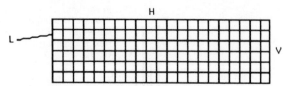

FIGURE 5.6   Rectangular lattice of spacing L, on which an atom carries out a random walk in the H and V directions.

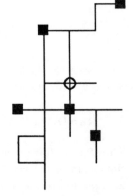

FIGURE 5.7   Example of random walks on a lattice in which the walker was permitted to return to points that had been previously visited. Tracks and endpoints (squares) of five walks of 10 steps each are shown. The starting point is circled, and, in this case, the probabilities were equal in the four possible directions.

If the walker has no memory of previous steps, motion in the horizontal direction will be independent of vertical motion. This independence means that we can treat motion in each direction as a statistically independent random walk. Thus the x-distribution and the y-distribution of endpoints can each be described by a histogram of the type in Figure 5.5b, corresponding to one-dimensional walks. Quantitatively, the result of this statistical independence of the motions in the two directions may be described as follows:

(1) Let $P_x(N, m)$ be the probability of a net displacement of m steps in the x-direction after a walk of N steps. It is given by equation (4.5) in the previous chapter. (This is the Bernoulli distribution.)

(2) Similarly, $P_y(N, n)$ will be the probability distribution for a displacement of n steps in the y-direction.

(3) The joint probability of a displacement of m steps in the x-direction and n steps in the y-direction is the product $P_x(N, m)P_y(N, n)$. This is the probability of moving to the point (m, n) in N steps.

We may get a qualitative sense of the motion if we suppose, for the moment, that the atom is confined to move only horizontally. The horizontal spread, h, of the atom's location will increase as the square root of the number of steps: $h \equiv \sqrt{\langle D^2 \rangle} = L\sqrt{N}$. When the atom is permitted to move in both directions, one-half of the steps, on the average, will be vertical and will not alter the atom's horizontal position. Hence, for the two-dimensional walk, the added dimension will increase by a factor 2 the number of steps required for an expected net displacement *in a particular direction.*

That is, the horizontal and vertical spreads are each given by $L\sqrt{(N/2)}$, and the mean square net displacement increases as $\langle R^2 \rangle = L^2 N$. So the rms radial distance moved on a plane surface in N steps of length L is just

$$\sqrt{\langle R^2 \rangle} = L\sqrt{N}. \tag{5.9}$$

This is identical to the expression for the spread of points in a one-dimensional walk. Hence the rms net displacement is not affected by the dimensionality of the space in which the walk takes place. This result is easily extended to higher dimensions.

Exercise 5.7
Consider a flat crystal represented by a two-dimensional lattice. Suppose an atom is inserted at the center of a circle inscribed with a radius of 6 lattice spacings. How many steps will be required, on the average, for the atom to wander outside the circle? Verify your solution by carrying out a two-dimensional random walk with a computer.

We may generalize the results of this section as follows:

After N steps of a random walk on an r-dimensional lattice,
the mean square of the net number of steps along a particular
direction is given by

$$\langle D^2 \rangle = \frac{L^2 N}{r}. \tag{5.10}$$

The expected square of the net *radial* distance $\langle R^2 \rangle$ is the sum of squares of the distance along the r-coordinate axes. As a consequence of the Pythagorean theorem, we have

$$\langle R^2 \rangle = r \langle D^2 \rangle = L^2 N. \tag{5.11}$$

This leads to the remarkable result that the dimensionality of the space, r, does not appear in the final expression for the net displacement in terms of N and L.

We are now in a position to discuss, in a crude fashion, the escape of a photon from a cloud or a fog bank (see Figure 5.1). Suppose a photon is created near the center and then moves in random directions with steps of fixed length L.

Figure 5.8 depicts the result of a simplified calculation in which the cloud is a disk and the photons perform a two-dimensional random walk from the center until they arrive at the surface. The times of these arrivals are indicated with tick marks. As we would expect, escape from the smaller cloud (below) is quicker.

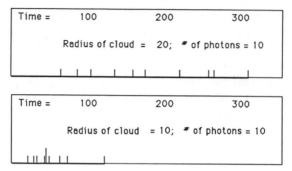

FIGURE 5.8 Times of escape of 10 photons emitted at the centers of circular two-dimensional clouds. Each tick on the horizontal axis is the escape of a photon, and the events occur earlier in the smaller star.

We may estimate the escape times as follows. (In Chapter 6, we do a more complete job of accounting for the surface of the cloud.) We ask how many steps will be required for the dispersion of the photon's position to equal the cloud's radius? When this number has been accumulated, it is a fair bet that the photon will have escaped from the cloud's surface. The ratio $m = R/L$ is the cloud's radius in units of the mean step-length. If the photon continues its random walk for $N = m^2$ steps, we expect it to be near the surface. Thus we find for the required length of the random walk to escape the cloud that

$$\langle N \rangle = \frac{R^2}{L^2}. \tag{5.12}$$

The quantitative consequences of this formula are rather striking, as the next exercise shows.

Exercise 5.8
The radius of the sun is 7×10^{10} cm, and taking $L = 1$ cm as a typical value for the step-length of photons inside the sun, find the number of steps and the total path-length required for a photon to escape from the center. At the speed of light, how long will this walk take? If the photon were moving in a straight line, how far across the Milky Way could it travel in this time?

Exercise 5.9
Consider an asymmetric walk on a two-dimensional, flat lattice, in which the probability of an atom's moving horizontally is twice as great as the probability of moving vertically. How does the mean shape of the pattern depend on the ratio of probabilities?

Exercise 5.10
Consider an asymmetric walk on a two-dimensional, flat lattice, in which the probability of an atom's moving to the right is twice as great as the probability of moving to the

left. How does the ratio of probabilities affect the atom's behavior? Does the mean shape of the pattern depend on the ratio of probabilities?

Exercise 5.11
Consider a walk on a two-dimensional lattice in which the horizontal spacing is much greater than in the vertical. How does the root-mean-square displacement increase with the step-length N?

(b) Triangular Lattice

Figure 5.9 illustrates a triangular lattice in which each node is numbered and represents a possible state of the system. We may interpret the node as a physical location in a two-dimensional space, and the lines represent pathways or "traffic lanes" between the pairs of adjacent locations. On the other hand, we may suppose the nodes represent abstract states, and the paths represent transitions between these states. The number of paths coming into a node will be the "path number" of that node.

Suppose we carry out a series of walks on a large example of such a lattice, limiting each walk to N steps. By "large" we mean that no edges are within N steps of the initial position of the walker. We will find that the radius of the cloud of endpoints increases with N in much the same way as it does on a rectangular lattice. The next exercise is a study of this behavior.

Exercise 5.12
Investigate the average net displacement of random walks of N steps on a large triangular lattice. How does the rms displacement depend on number of steps when the usual Pythagorean theorem does not apply?

But now let us see what happens if we carry out a long walk (10,000 steps) on the lattice of Figure 5.9. In this case, the walker is trapped and has many

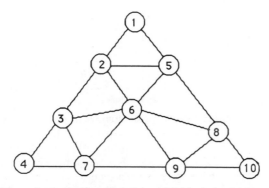

FIGURE 5.9   Triangular lattice in which the nodes (circles) are connected by a varying number of paths to neighbors. Node 6, for example, is said to have a path number of 6. We seek the expected number of visitations to various nodes and we find that the result depends on the method of averaging.

chances to visit each node. We wish to find the average state of the system, that is, the fraction of times we would expect to find the system in each node if we were to take a series of snapshots. The result will depend on the method by which we carry out the walk.

Suppose, for a start, we do it as follows. At each tick of a fictitious clock, the walker chooses among a set of equally likely paths. It either selects a path leading away from its present location or it selects a null path that keeps it at the same location. If the walker remains at the same location, we say it underwent a "null transition." Thus, for example, a walker in one of the nodes from which there are four paths leading to other nodes has a probability $p = 1/5$ of going to each of the adjacent nodes and a similar probability of remaining it its current location. See Figure 5.10 (upper) and see Section 7.7 for a more detailed discussion of the computation of the walk.

We find that the traffic is about the same on all paths, and nodes with fewer paths leading to them are less often visited. In fact, the number of times we find the system at each node is proportional to the path number of that node.

But there are other ways to carry out this walk.

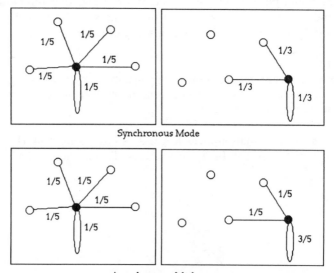

FIGURE 5.10   Two modes of calculating random walks on a two-dimensional lattice. The solid dot indicates the occupied node at a particular time, and the lines indicate possible paths to neighbors. The loops are null transitions that leave the system in its starting point. The modes differ in the way they assign probabilities to the paths. In both cases the sum of probabilities from a node is unity, so a transition occurs in every time-step. In the synchronous mode (upper) *all paths from a node* have the same probability, while in the asynchronous mode (lower) *the exit paths of all nodes* have the same probability. In the asynchronous mode, nodes with many paths are exited more quickly.

5.7   THE MODES OF A RANDOM WALK IN TIME

When we evaluate the time-averaged behavior of such a walk, we must specify the method of averaging. This method of averaging is called the "mode" of the walk, and it is related to the method of carrying the walk forward in time.

The mode used in the preceding section is one of many possibilities, and we call it the "synchronous" mode. The probability of leaving a node, $p = (n - 1)/n$, is nearly independent of the path number n, as long as n is not too small. Thus the walker moves with a fixed rhythm. We found that all paths, including the null parths, were equally busy and the relative population of each node was proportional to the number of paths leading to it.

Another possible method of carrying out the walk is called the "asynchronous" mode, and it is illustrated in Figure 5.10 (lower). In this mode, each path contributes an equal probability of making a transition in one click of the computer clock. Thus the walker leaves more quickly from nodes with large path numbers. As before, the traffic along various paths are equal. But if we take snapshots of the system at random (or uniform) intervals of time, we find a different appearance. In this mode of calculation, all nodes are populated with the same average frequency.

We return to a discussion of this second mode in Chapter 7, where we represent the shuffling of energy among three atoms as a random walk on a triangular lattice.

5.8   RANDOM WALK IN VELOCITY SPACE: BROWNIAN MOTION

We now attempt to build a mathematical model for the irregular motion of small particles (such as dust motes) in a gas or liquid (Brownian motion). Rather than assume that the *position* of the particle can jump from one value to another, it is more natural to suppose that the *velocity* can behave this way. This will lead us to a new type of random walk.

We can imagine that the mote undergoes an abrupt change of velocity each time it is hit by a molecule. (The larger the particle, the smaller is the change produced by each molecular impact.) If we confine ourselves to one-dimensional motion, we might describe the successive positions and velocities along the x-axis by the following pair of equations:

$$x_{i+1} = x_i + v_i \delta t, \tag{5.13a}$$

$$v_i = v_{i-1} + a_{i-1} \delta t. \tag{5.13b}$$

The term a_{i-1} is the acceleration of the particle, caused by collisions with atoms. In Chapter 13, we examine this term in more detail, and for the moment we borrow from those results to say that the acceleration term can be approximated by breaking it into two parts: (1) a rapidly fluctuating term accounting for the

TABLE 5.1 Spreadsheet Program for Solving the Three Equations Governing the Random Walk with Friction

	A	B	C	D	E	F	G	H
1		x	v	a	eps	alpha	dt	eps0
2						2	0.05	0.5
3								
4		1	0.1	=-$F$2*C3+E3	=(RAND()-0.5)*$H$2			
5		=B4+C4*$G$2	=C4+D4*$G$2	=-$F$2*C4+E4	=(RAND()-0.5)*$H$2	1	=B5-1	=C5
6		=B5+C5*$G$2	=C5+D5*$G$2	=$F$2*C5+E5	=(RAND()-0.5)*$H$2	2	=B6-1	=C6
7		=B6+C6*$G$2	=C6+D6*$G$2	=-$F$2*C6+E6	=(RAND()-0.5)*$H$2	3	=B7-1	=C7
8		=B7+C7*$G$2	=C7+D7*$G$2	=$F$2*C7+E7	=(RAND()-0.5)*$H$2	4	=B8-1	=C8

individual collisions with neighboring atoms, and (2) a much slower term that behaves like friction. The friction term results from the fact that as the particle moves through the gas, it is more often hit on the front face than on the back face—just as a runner in the rain is likely to get wetter on the face than on the back. This asymmetry tends to slow down the particle, so it acts like friction. The faster the particle, the greater the effect.

So let us write one more equation to describe the behavior of the Brownian particle, as $a_{i-1} = -\alpha v_{i-1} + e_{i-1}$, where α is the friction coefficient, and e_{i-1} represents the effects of individual collisions. With elementary calculus, it is possible to show that these three equations are equivalent to the second-order differential equation,

$$\frac{d^2x}{dt^2} = -\alpha\frac{dx}{dt} + e. \tag{5.14}$$

This equation (known as the Langevin equation) is Newton's law of motion for an object of unit mass under the influence of friction and a random external force e. We shall, however, leave the solution of the differential equation to the interested reader and shall instead use a spreadsheet program to solve the original set of difference equations. Table 5.1 shows the equations in the spreadsheet program, and each column (B–E, respectively) is devoted to one of the variables x, v, a, and e. The value of e is computed from the random-number generator, and the values of each variable are computed from the values in the preceding row in conformity with the equations.

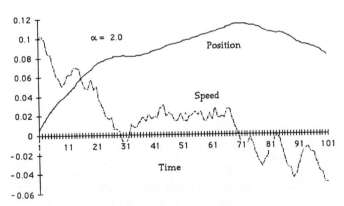

FIGURE 5.11a Solution of the equations for the random walk with friction ($\alpha = 2.0$, $|e| = 0.5$) obtained with the spreadsheet program in Table 5.1. Time increases to the right and the curves show the particle speed (dashed) and position on the x-axis. The time span is 100 steps, and a collision with an atom occurs at each time step. Note that the speed curve is much more jagged and that the position curve shows primarily slow variations. In other words, the observed displacements (Brownian motion) of small particles do not reveal individual collisions but the effects of groups of collisions.

Figures 5.11a and 5.11b show solutions that span 100 time-steps and assume a collision to occur at every time-step. Note two features of these solutions. First, the velocity (dashed line) is much more irregular and jagged than the position. The velocity shows the effects of individual collisions, while the position does not. Second, the velocity variations occur on two distinct time scales: (1) a short one corresponding to the fact that a particle hits at every time-step and (2) a long one corresponding to the effect of friction and the accumulation of many collisions. A mathematical study of the equation shows that the position shows two time scales also. On the short time scale, the particle moves with a nearly constant speed, so its departure from its initial position increases with time as $r = vt$. This can be seen at the start of the curve, for example, where the slope of the position curve remains nearly constant for a while. On the longer time scale, the particle moves as though it were doing a random walk, so its departure increases only as the square root of the time. The combined effect of friction and the random collisions is to make the particle gradually lose information about its original velocity.

Figures 5.11a and 5.11b show random walks with different amounts of friction, represented by the factor α in the differential equation describing the motion of the particle under the influence of random impacts. The difference in the influence of the two values of α can also be seen in Figure 5.12, where the response of the Brownian particle to a single impulse, rather than a random train of impulses, is shown. The larger value of friction corresponds to the more rapid decrease of the speed, so the fluctuations are somewhat smaller and on a shorter time scale with the larger friction. The effect of the fluctuating force is dissipated more rapidly with the larger value of α.

The fact that larger friction leads to smaller overall fluctuations (given the same impulses) is one aspect of the "fluctuation–dissipation" theorem. This

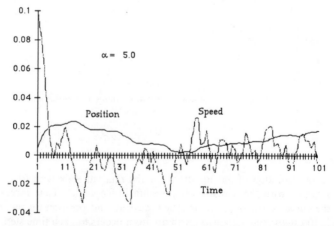

FIGURE 5.11b Similar to Figure 5.11a with larger friction ($\alpha = 5.0$). The effect of the larger damping is to make the particle return to zero speed more quickly, so the fluctuations are slightly smaller.

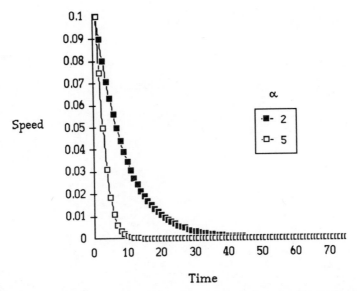

FIGURE 5.12   Response of Brownian particle to a single impulse, illustrating the effect of varying the amount of friction. The two curves show the speed as a function of time for two values of α. These are the same values as used in Figures 5.11a and 5.11b.

theorem provides a relationship between the fluctuations experienced by Brownian particles and the dissipation of energy in the gas. We discuss another aspect of this theorem in Chapter 13, where we show how the amount of friction felt by a Brownian particle can be related to the nature of the particle and the atoms with which it collides.

REFERENCES

Chowdhury and Mookerjee 1985
Hersch and Griego 1969
Kruglak 1987
Mishima et al. 1980
Mohling 1982
Schumacher 1986
Wax 1954

6

RANDOM WALKS NEAR BOUNDARIES

Chapter 5 demonstrated that, if a large number of walks start simultaneously, the radius of the diffuse "cloud" of walkers will increase as the square root of the number of steps—or, equivalently, the square root of the time. An approximate solution to the escape problems posed in Chapter 4 (for the frog from a pipe or an atom from a crystal) was achieved by ignoring the effect of the boundary. We carry out the calculation as though the space were unlimited and evaluate the number of steps required for the cloud of walkers to expand so that it reaches the edges of the space. In other words, the boundaries were ignored in constructing the path of each random walk. Such calculations only give rough estimates because they do not explictly consider the influence of the edges on the shape of the cloud.

6.1 PHOTONS ESCAPING FROM A CLOUD

(a) Sample Calculations

Figure 6.1 illustrates a method of testing this approximation for a two-dimensional walk of photons near the surface of a cloud. In the top frame is shown two sample walks of 30 steps of randomly chosen length and direction (similar to the walks in Figure 5.1). The walks start at the ×, and the vertical line is the free surface. One of the walks has penetrated the free surface, so the photon flies off into space and the walk terminates. The middle frame shows the results of 70 such walks, although the tracks are not shown. The 58 dots are the endpoints of walks that did not escape the cloud, and we note that their

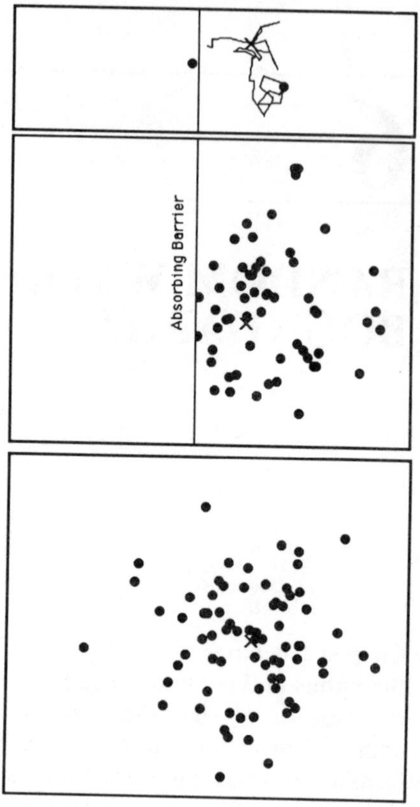

FIGURE 6.1 Random walks in near the edge of a cloud (vertical line), from which photons may escape. The purpose of this set of calculations is to test estimates based on calculations that did not explicitly account for the barrier. The top frame shows the tracks of two walks; the middle frame shows the endpoints of 58 walks (out of 70) that did not escape; the bottom frame shows the endpoints of 70 walks computed as though the cloud were infinite. Only 10 photons ended to the left of the edge. (See text.)

center of gravity is displaced away from the surface. Twelve photons out of the 70 have escaped through the surface. The lower frame shows the result of 70 walks in the absence of a free surface.

To use such a calculation to estimate the effect of a free surface, we count the number of photons that ended at points beyond the position of the surface. According to this measure, only 10 would have escaped. This is too small a sample for generalizing, but the result is in qualitative agreement with what we might expect. The reasoning is the following. In the middle frame, none of the photons that ended inside the cloud (to the right of the edge indicated by the vertical line) had ever been outside the cloud. On the other hand, in the bottom frame, some of the endpoints may correspond to points that had wandered to the left and then back to the right side of the edge. Hence we expect to find more endpoints that lie to the right of the edge in the bottom frame than in the middle frame. Thus ignoring the effect of the absorbing boundary tends to under estimate the expansion of the cloud of endpoints.

Thus the approximation that evaluates the influence of the edge of the cloud by mapping the walks as though the cloud were infinitely large will tend to under estimate the number of photons that are lost from the cloud.

In the next section, we approach the solution of the escape problem by treating a one-dimensional space that is bounded at one end. Such a boundary is "absorbing" or "reflecting" depending on whether the walker is lost at the boundary (and the walk terminates) or is deflected back into the half-space. An atom (or the frog in a pipe) is said to have met an absorbing barrier if it comes to the open end and jumps out, finishing the walk. A reflecting boundary would consist of a closed end, where the atom would bounce back and continue hopping inside.

Exercise 6.1
Both of these parts may be done with one-dimensional random walks.

(a) Carry out a series of random walks with and without an absorbing barrier to test the rule stated above, that the escapes are underestimated if we do not explicitly account for the barrier.

(b) What happens when the barrier is a mirror? Will the approximation overestimate or underestimate the number of photons that strike the mirror in a walk of N steps? Make a prediction and then modify your random-walk program to test your conclusion.

(b) One-dimensional Approximation

We consider a thin region near the surface of a cloud, represented by a flat layer. Photons produced inside the cloud execute random walks toward and away from the surface, and we assume that the photons are merely scattered by the gas, so they continue to walk indefinitely. Some of the photons quickly emerge from the cloud and escape into space, while others move into the cloud before wandering to the surface. Figure 6.2 illustrates the pattern of emergence

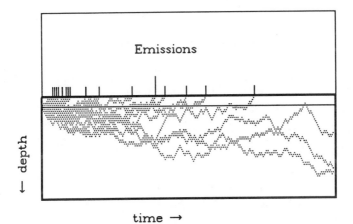

FIGURE 6.2 Random walks of 20 photons created at a distance of 5 steps from the surface of a cloud. Vertical ticks show the times of emission from the surface. Three photons remain in the cloud at end ($t = 300$).

when 20 photons are produced simultaneously at a distance of 5 steps from the surface. It is similar to Figure 5.10 except that it relates to a flat, rather than a round, region, and it is therefore easier to analyze. Our aim in this chapter is to understand this pattern of emergence.

Exercise 6.2

Construct a computer program to carry out simulations illustrated in Figure 6.2. Initiate a group of *n* photons at a depth *D*, and let them execute random walks upward and downward along the perpendicular to the surface. When a photon reaches the surface, tabulate an escape and prepare a histogram like Figure 6.2, showing the times of escape. Note that the frequency of emissions (the "light curve") has an abrupt rise when the photons are injected near the surface (*D* is small). How does the shape alter when *D* is increased?

Exercise 6.3

Using the program developed in Exercise 6.2, show that the mean distance of the photons from the surface increases with time, because the surface acts as a limitation to the distance in one direction, while there is no corresponding limitation to distance in the other direction. Can you concoct an approximate relationship for the rate of increase? Test it.

Two idealized types of boundary, reflecting and absorbing, have quite distinct effects:

(1) An absorbing barrier or a free surface *decreases* the probability that the photon will find itself at a particular point in the interior of a cloud, because the paths that would have gone beyond the barrier and returned to the point have been eliminated.

(2) A reflecting barrier, such as a mirror, *increases* the probability of being at a point because many paths have been turned back into the half-space.

TABLE 6.1   Endpoints of 1000 Random Walks near Absorbing Boundary

N	0	1	2	3	4	5	6	7	8	9	10	11	12	13	Number of Survivors
0				1000											1000
1			494	0	506										1000
2		249	0	494	0	257									1000
3		0	268	0	354	0	129								751
4		125	0	318	0	246	0	62							751
5		0	152	0	281	0	161	0	32						626
6		75	0	200	0	240	0	90	0	21					626
7		0	87	0	229	0	176	0	51	0	8				551
8		49	0	143	0	212	0	110	0	36	0	1			551
9		0	69	0	166	0	177	0	67	0	23	0	0		502
10		32	0	120	0	170	0	130	0	42	0	8	0	0	502

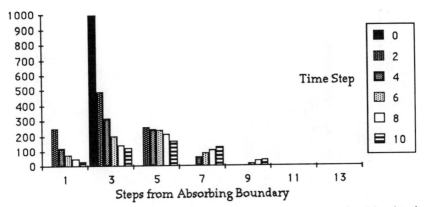

FIGURE 6.3 Frequency of visits for 1000 random walks near an absorbing barrier, based on Table 6.1.

Table 6.1 and Figure 6.3 illustrate a series of walks that start 2 steps from an absorbing barrier. The numbers at each epoch, N, can be considered as the distribution of endpoints for walks of length N. (Alternately, they may be considered as the number of visits to each point during the process of 1000 walks.) The number of "survivors" at each epoch indicates the number of walkers that have not yet touched the absorbing boundary.

Table 6.2 and Figure 6.4 show the results of a series of one-dimensional random walks in the presence of a reflecting boundary. In this case, all walkers survive, and they are merely turned back at the boundary. In the next section, we develop a technique for predicting these patterns.

TABLE 6.2   Endpoints of 1000 Walks near Reflecting Barrier

					Distance from Barrier									Number of
0	1	2	3	4	5	6	7	8	9	10	11	12	13	Survivors
N														
0			1000											1000
1		495	0	505										1000
2	257	0	472	0	271									1000
3	0	496	0	376	0	128								1000
4	261	0	426	0	252	0	61							1000
5	0	475	0	335	0	155	0	35						1000
6	240	0	408	0	250	0	83	0	19					1000
7	0	435	0	340	0	165	0	51	0	9				1000
8	212	0	391	0	253	0	112	0	28	0	4			1000
9	0	405	0	330	0	174	0	77	0	11	0	3		1000
10	202	0	371	0	250	0	125	0	44	0	8	0	0	1000

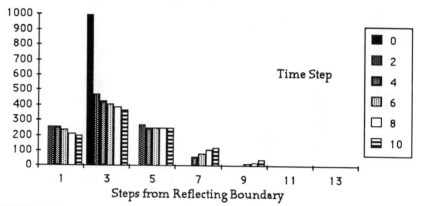

FIGURE 6.4 Frequency of visits for 1000 random walks near a reflecting barrier, based on Table 6.2.

FIGURE 6.5 Atom starts a random walk in a one-dimensional lattice at A. What is the chance that it will escape through the surface at B before visiting the site at C?

Exercise 6.4

Modify the program developed in Exercise 6.2 so that photons are reflected away from the cloud surface rather than being emitted into space. This produces a series of random walks in the presence of a reflecting boundary. Compare the results with the previous case. Before reading ahead, see if you can derive a scheme for assessing the influence of a single boundary at a certain distance from the starting point.

Suppose an atom is wandering back and forth on a one-dimensional lattice, as in Figure 6.5. The left end of the lattice is open (B), and there is a special lattice site at C. The atom starts from point A, and we ask the probability that it will reach the surface at B without first visiting the site at C. The reflection theorem provides the answer.

6.2 THE FELLER REFLECTION THEOREM

A general principle for solving such a random walk problem is the "reflection theorem," whose proof was first given by W. Feller (1968). The power of the theorem is remarkable, and its consequences are often surprising.

In Figure 6.6, we represent the photon's walk in a schematic space–time diagram in which space is plotted vertically and step-number (time) is plotted horizontally along the t-axis. $A = (0, a)$ is the starting point, and $B = (N, b)$ is an arbitrary point N time-steps along the way. At each time-step, the walker advances to the right, along the time-line. Its spatial motion is given by the

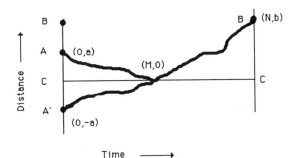

FIGURE 6.6 Schematic plot of one-dimensional random walk. Time is measured horizontally and distance vertically. The walk starts at A, which is a distance a from the line. The reflected point, A', is at a distance $-a$ from the line.

random variable $\delta y = +1$ or -1, and this is plotted in the y-direction. Thus, at each time step, $y \to y + 1$ or $y \to y - 1$.

A "direct" path from A to B is a series of steps $\{Y_1, Y_2, \ldots, Y_N\}$ leading from A to B. The "reflection" of A is designated A', and it is the point at the same epoch as A, but with $a = -a$. (It is similar to the mirror image of A, so A and A' are equidistant from the x-axis.)

The paths from A to B may be divided into two categories. Some paths $A \to B$ cross the x-axis and others do not. The "reflection principle" describes the number of paths from A to B that touch or cross the x-axis. It may be stated as follows:

Reflection Theorem: *The number of paths from A to B that either touch or cross the x-axis equals the total number of paths from the reflected point A' to B.*

The proof is as follows. Referring to Figure 6.6, draw an arbitrary path $A \to B$ which touches the x-axis for the first time at the point M. Construct the reflected path $A' \to M$. For each path $A \to M$, there is exactly one reflected path, $A' \to M$. Thus there is a one-to-one correspondence between the paths $A \to M \to B$ and $A' \to M \to B$ which have the section $M \to B$ in common. This implies that there is a one-to-one correspondence between paths $A \to B$ and paths $A' \to B$ that touch the x-axis at M. But M is an arbitrary point on the axis. Thus there are

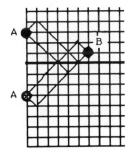

FIGURE 6.7 According to the Feller reflection theorem, the number of paths $A \to B$ that move along diagonals and touch or cross the heavy line equals the number of paths $A' \to B$ from the reflected point. In this case there are three of each. (A and A' are equidistant from the line.)

as many paths $A \rightarrow B$ that touch or cross the x-axis as there are paths from $A' \rightarrow B$, all of which must cross the x-axis. This completes the proof.

Exercise 6.5
Referring to Figure 6.7, verify the reflection theorem for the paths $A \rightarrow B$ by counting the legitimate paths that touch or cross $a = 0$ and show it equals the number of paths from the reflected point $A' \rightarrow B$.

6.3   INFLUENCE OF AN ABSORBING BARRIER

As an application of the reflection principle, suppose the site C (Figure 6.5) is a "hole" that can capture the wandering atom. What does this do to the atom's probability of arriving at B and escaping from the surface? Clearly, it reduces the chance, but how much?

To solve this problem, we need to evaluate the probability of walking from A to B, undergoing a net displacement m, without touching the absorbing barrier (Figure 6.8).

We need to count the number of paths from A to B that do not touch the x-axis, and this is the *total* number of paths from A to B *less* the number that touch or cross the x-axis. The *total* number of paths in a walk of length N leading to a net displacement of m is $W_{N,m}$, defined in Chapter 4, and this is the total number connecting points A and B. From Chapter 4 we have equation (4.6):

$$W_{N,m} = \frac{N!}{[(N + m)/2]![(N - m)/2]!}$$

Furthermore, the number that touch the barrier on their way to B is the number of paths from the reflected starting point $(A' \rightarrow B)$, and this is $W_{N,2a+m}$. (Note that reaching point B from the reflected starting point requires a

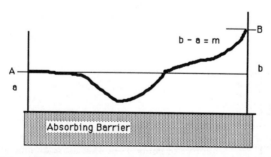

FIGURE 6.8   Schematic representation of atom's random walk in Figure 6.5. The note at C is depicted as an absorbing barrier.

displacement $2a + m$.) Thus the number of paths $A \to B$ that do *not* touch the x-axis is the difference, or $W_{N,m} - W_{N,2a+m}$. We may write the probability of the corresponding path as

$$P_{abs}(N, m; a) = [W_{N,m} - W_{N,2a+m}][1/2]^N. \qquad (6.1)$$

This is the desired solution, and it describes the expected entries in Table 6.1.

Exercise 6.6
Compare expression (6.1) with the randomly generated numbers in Table 6.1.

6.4   INFLUENCE OF A REFLECTING BARRIER

When the walker arrives at a reflecting barrier, it has unit probability of turning back, instead of 1/2. Thus the expected number of paths leading to B from that point of the barrier is doubled by that reflection. Additional reflections along the same path will lead to additional factors of 2. We may achieve this increase by adding the number of paths $A' \to B$ to the number of paths $A \to B$. Hence we have for the number of paths from A to B in the presence of a reflecting barrier, $W_{N,m} + W_{N,2a+m}$, and the corresponding probability is

$$P_{ref}(N, m; a) = [W_{N,m} + W_{N,2a+m}][1/2]^N. \qquad (6.2)$$

Table 6.2 shows the results of 1000 walks starting at a distance of 2 steps from a reflecting barrier.

Exercise 6.7
Compare this expression with the randomly generated numbers in Table 6.2.

6.5   THE BALLOT THEOREM

A problem whose solution has many useful applications may be stated as follows: What is the probability of walking from the origin to point (N, m) along a path for which $Y_i > 0$, that is, along a path that never returns to the x-axis?

This is called the "ballot problem" because it is equivalent to asking for the likelihood that the winner of a two-candidate election was always in the lead if the final count of N votes indicates a margin, m. (The counting of the ballots may be treated as a random walk with steps $+1$ and -1 for a particular candidate.)

Referring to Figure 6.9, the total number of paths from the origin to (N, m) is $W_{N,m}$. The paths we wish to count are the paths from $(1, 1)$ that reach (N, m) without returning to the x-axis. The number of such paths is given by

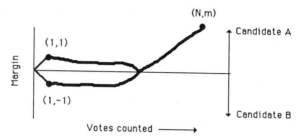

FIGURE 6.9   The ballot problem. A total of N votes are counted for two candidates, and the running margin of A over B is plotted vertically. What is the probability that the winner, A, will be ahead throughout the counting if A's final margin is m?

$$
\begin{aligned}
W_{N-1,m-1} - W_{N-1,m+1} &= C(N-1,[N-1+m-1]/2) \\
&\quad - C(N-1,[N-1+m+1]/2) \\
&= C(N-1,n-1) - C(N-1,n). \\
&= \left(\frac{m}{N}\right) W_{N,m}
\end{aligned}
\tag{6.3}
$$

(The reader should prove the last step.) Dividing this by the total number of paths to the point (N, m), we find that the fraction of paths to the point (N, m) that do not return to the x-axis is simply m/N. We may express the result as follows:

Ballot Theorem: *In counting N votes between two candidates, the probability that the eventual winner will never have had fewer counted votes than the loser is equal to the final margin m divided by N.*

For example, if one candidate eventually wins by a ten percent margin, there was a ten percent probability of that candidate's being ahead throughout the counting process.

An important corollary has to do with the probability of a walk's returning to the origin. We know that the expected displacement of a random walk of N steps of unit length is $\sqrt{N}$. Suppose such a walk achieves exactly the expected displacement. According to the ballot theorem, the probability that it will achieve its expected displacement *without* returning to the origin is $\sqrt{N}/N = \sqrt{1/N}$. For a long walk, this probability becomes small, and as the walk-length increases without limit, the probability that it *will* return to the starting point approaches unity.

This also implies that the photons of Figure 6.2 that move into the cloud will return to the surface, given sufficient time and a cloud that is infinitely deep in one direction.

6.6 APPLICATION OF THE BALLOT THEOREM: ARRIVALS AT A BARRIER

The ballot theorem puts us in a position to analyze the results illustrated in Figure 6.2. Specifically, we can answer the following question: If a group of photons is generated simultaneously at a particular distance from the edge of a cloud and execute one-dimensional random walks, what will be the distribution of arrival times at the surface? In other words, how will the emission of photons from the surface vary with time?

The surface is an absorbing boundary, because photons never return to the interior once they have arrived at the surface. (Although the photon is not physically absorbed, it is lost to the cloud.)

Figure 6.10 illustrates the problem. We ask for the probability of arriving at (N, m) without having previously touched the line at m.

We phrase the question as follows: If a particle starts at the origin, what is the probability that it will strike an absorbing boundary m for the first time at the epoch N? If we assume each step occupies the same interval of time, the walk-length N serves as a measure of lapsed time.

To solve the problem, we first rephrase it. There are as many paths $A \rightarrow B$ as there are $B \rightarrow A$. Thus we can consider the reversed path and ask: What is the probability $Q(N, m)$ of achieving a displacement m in a walk of length N without first returning to the origin? But this is just the balloting question, and the required probability is just the total probability of getting to the point m after N steps, $P(N, m)$, times the fraction of all paths that do not touch the x-axis. Thus it is

$$Q(N, m) = P(N, m)[m/N]. \tag{6.4}$$

With $P(N, m) = W_{N,m}[1/2]^N$, we have

$$Q(N, m) = W_{N,m}[1/2]^N[m/N]. \tag{6.5}$$

As an example, consider the walks illustrated in Figure 6.2, which started at a point that is $m = 5$ steps from a free surface, which acts like an absorbing boundary. What are the probabilities of arriving at the surface after $N = 5, 6, \ldots$ steps? (In this case N cannot be an odd number.)

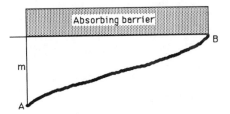

FIGURE 6.10 Schematic representation of random walk that ends on an absorbing barrier.

TABLE 6.3   Probability of Absorption at $m = 5$ after Exactly N Steps

N	$5/N$	$P(N, 5)$	$Q(N, 5)$
5	1	0.03125	0.03125
7	5/7	0.05469	0.03906
9	5/9	0.07031	0.03906
11	5/11	0.08057	0.03662
13	5/13	0.08728	0.03357
15	1/3	0.09164	0.03055

TABLE 6.4   Arrivals at Absorbing Barrier, $m = 5$, in 1000 Walks

N	Number of Survivors	Number of Absorptions	Expected Number
3	1000	0	0
5	971	29	31.2
7	936	35	39.1
9	902	34	39.1
11	863	39	36.6
13	822	41	33.6
15	776	46	30.5
17	749	27	27.8
19	717	32	25.3
21	688	29	23.1
23	664	24	21.2
25	646	18	19.5
27	630	16	18.0
29	616	14	16.7

The resulting probability $Q(N, 5)$ is shown in Table 6.3. It has a maximum near $N = 7$–9. Table 6.4 shows the detailed calculation for the early portion of 1000 random walks that started at a distance of 5 steps from the boundary. The first and second columns give the current length of the walk (epoch) and the number of walkers that have not struck the boundary. The third and fourth columns give the computed and expected number of arrivals at the boundary at each epoch. Both sets of data show the initial rise and subsequent decline in the rate of arrivals at the boundary, and the agreement is good.

Exercise 6.8

Compare the calculated arrival frequencies in Table 6.4 with the results of the simulation performed in Exercise 6.1.

6.7   RANDOM WALK ON A FINITE LATTICE: AVERAGE SOLUTION FOR THE LONG RUN

(a) Equal Transition Probabilities

Suppose the walker finds itself on a small linear lattice, as illustrated in Figure 6.11. Both boundaries are reflecting barriers, so the walker is trapped. The jumping rule is that at each time-step the walker must move one space to the right or to the left. The probability of jumping right is P_R, and the probability of jumping left is $1 - P_R = P_L$. We assume the walk continues indefinitely, returning time and again to each site. As time goes on, the number of visits to each site becomes a well-defined fraction of the total visits to all sites. How does this fraction vary from site to site? This is the subject of Exercise 6.9.

Exercise 6.9
Carry out a random walk on a linear lattice in which the endpoints are reflecting barriers. Keep track of the number of visits to each site. Assume equal probabilities of moving to the right or left, and show that the frequency of visits to any site is proportional to the number of possible routes into that site. Thus the endpoints have half as many visits as the interior points.

Now suppose the lattice is a two-dimensional graph of nodes and connections, as in Figure 6.12. We characterize each node by its "connection-number," the number of lines leading to it from other nodes, and we adopt the following jump rule: At each time step, the walker must randomly select one of the exits from its current node and then follow it to the new node. We assume that the exits from a given node all have equal chance of being selected. This rule implies that the walker will move with a constant rhythm from node to node.

i   i+1

FIGURE 6.11   A linear lattice in which an atom is trapped and executes an endless random walk. The sites are indexed by i, increasing to the right.

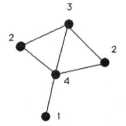

FIGURE 6.12   Graph of nodes and connections on which a random walk takes place. Each node is marked with its connection-number, indicating the number of lines leading to and away from that node. We seek the relationship between number of vists and connection-number with the jump rules defined in the text.

We keep track of the number of visits to each node, and our question is: How will the relative number of visits to a node depend on the connection-number of that node?

Exercise 6.10
Carry out a walk on a graph such as Figure 6.12 and show that the visitation number is proportional to the connection-number when the transition probabilities from any given node are equally divided.

An important variation on this calculation (which we do not explore) is the following. Suppose we change the jump rule so that the walker leaves a node more quickly if there are more connections out of that node. We watch for a long time and we ask not only for the relative *numbers of visits* to each node but also for the relative *amount of time* it spends at each node. The answers to these two questions are not the same. Thus it is crucial to define carefully the transition rules and the nature of the averaging process when we come to apply these models to physical systems. (See Section 5.7.)

(b) Unequal Transition Probabilities

Now let us return to the simple lattice in Figure 6.11 and assume the probability of jumping right is greater than the probability of jumping left. What do we expect for the pattern of visitations in this case?

We may formalize the process as follows. Let N_i and N_{i+1} be the numbers of visitations to two adjacent sites in the one-dimensional lattice, with i increasing to the right. In the long run, the distribution of visitation numbers must be stationary and not change in shape. This can only occur if the number of jumps to the right between any pair of sites equals the number of jumps to the left between that same pair. (Or else the walkers would tend to flow in one direction and pile up at that end of the lattice.) This behavior is expressed as

$$\text{Rightward jumps } (i \rightarrow i + 1) = \text{Leftward jumps } (i \leftarrow i + 1)$$

or

$$N_i P_R = N_{i+1} P_L.$$

Thus the ratio of visitation numbers is given by

$$\frac{N_i}{N_{i+1}} = \frac{P_L}{P_R}, \tag{6.6}$$

for all i in the interior. The visitation numbers thus make up a geometric (or exponential) series, in which each term is a constant multiple of its neighbor. If $P_L < P_R$, the walkers tend to accumulate toward the right-hand end of the lattice. When they have accumulated sufficiently, the number of jumps will be the same in both direction.

Exercise 6.11
Set up a simple spreadsheet program to evaluate the relative numbers of visitations for the case $P_L/P_R = 1.05$. Plot the results on a logarithmic scale and show that they follow an exponential distribution.

REFERENCE

Feller 1968

7

SHUFFLING OF ENERGY
QUANTA IN ISOLATED SYSTEMS

In this and Chapter 8, we apply probabilistic ideas to highly simplified systems of atoms and energy quanta. These chapters provide an introduction to a wide and important branch of physics known as statistical physics.

7.1 THE AIM OF MODERN STATISTICAL PHYSICS

With a suitable choice of descriptive parameters, statistical physics and its older sister thermodynamics permit us to discuss systems that we cannot know—and do not wish to know—in complete detail. We do not attempt a detailed accounting; we confine attention to statistical parameters and distribution functions that are approached, in the average, more and more closely as time moves on. From these distribution functions, we may derive important properties of the gas, such as temperature, pressure, density, and stratification.

The primary distinction between the two sisters is that thermodynamics deals with measurable, macroscopic parameters, such as pressure and temperature. The results of thermodynamics are often quite general so they can be applied to a wide range of phenomena. For example, the "first law" of thermodynamics, which we describe later, is a general statement of energy conservation and it can be applied to the analysis of any type of engine, as long as the engine conforms to certain restrictions. Statistical physics, on the other hand, attempts to go deeper, by modeling the microscopic behavior of a system. The result is more powerful in some ways, and less powerful in other ways. Less powerful because statistical physics is more specific and cannot be applied with the sweeping generality of thermodynamics. More powerful because it does not impose such severe restrictions on the systems it can discuss.

Classical thermodynamics is, in fact, misnamed. It ought to be called "thermostatics," because it is confined to systems that are in equilibrium, which means that it only deals with systems that are isolated and not changing with time. But if a system changes slowly enough, it may be said to follow a concatenation of equilibrium states, and, in this way, thermodynamics can take slowly changing (quasi-equilibrium) systems under its wing. Having done so, it can tell us much about the system without detailed modeling. But this limitation to slow changes necessarily eliminates a broad class of important phenomena: the so-called "irreversible" changes. While a system undergoes such changes, thermodynamics must briefly close her eyes. She may reopen them when the system has settled down once more to an near-equilibrium state. One example is the conduction of heat in a gas. If we add energy to the gas at one end of a tube, some of this energy will be carried to the other end. If the heating is gentle and gradual, thermodynamics can describe the entire process; if the heating is drastic and abrupt, it cannot cope, and we must resort to statistical physics. We must proceed by constructing a model that permits us to specify the average behavior of the atoms in the gas.

Thus the range of phenomena open to statistical physics is far greater. This increased range is purchased at the expense of our needing to build a specific model, but computers can help us cope with such models.

Figure 7.1 schematically illustrates the question addressed by statistical physics that is the focus of this chapter. Suppose a specified amount of excitation energy is shared by a fixed number of atoms. The atoms exchange the energy by collisions and by emitting and absorbing photons. Suppose for the moment that each photon carries an equal quantum of energy, e. When an atom emits a photon, the atom's energy decreases by e, and when the photon is reabsorbed by another atom, that atom's energy increases by the same amount. For the moment, we ignore the atom's kinetic energy and we assume none of the photons can escape from the system, so the total energy is constant. Atoms may absorb more than one photon and, in doing so, they move into a higher level of excitation. According to this simplified model, the excitation energy of an atom

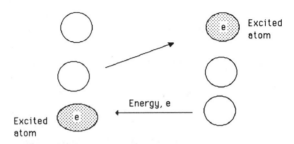

FIGURE 7.1   Schematic representation of atoms exchanging energy quanta. We seek the numbers of atoms in various amounts of energy.

at any moment consists of q quanta, and it equals qe. It is said to be in the qth state of excitation.

At a series of times we measure the excitation state of all the atoms—that is, we evaluate q for each atom—and we make a histogram showing the number of atoms in various states of excitation $n(q, t)$. The total excitation energy E is given by the sum $E = e\sum qn(q, t)$, and we assume this to be constant, independent of time.

The key question is: If we average this histogram over long periods of time, what will be the shape of the histogram. A "long period of time" will be an interval long enough so that the starting time or the ending time makes no difference in the average shape of the histogram. After a "long" time, every atom in the system will have engaged in, say, ten or more exchanges of quanta.

To be more specific, let us define $n_q \equiv \langle n(q, t) \rangle$ as the straight average of the distribution $n(q, t)$, counted at T times and added together,

$$n_q \equiv \langle n(q, t) \rangle = \frac{1}{T} \sum_{i=1}^{T} n(q, t_i). \tag{7.1}$$

According to the "ergodic hypothesis," a different type of averaging, in which we follow a particular atom, gives the same result. This hypothesis is discussed later in this chapter.

We are interested in the long-term effect of this random shuffling of energy. We call this random shuffling with constant energy the "Boltzmann process."

7.2 THE BOLTZMANN PROCESS WITH MARKERS ON A CHECKERBOARD

Suppose a number, Q, of markers are placed on the N squares of a checkerboard (Figure 7.2). These markers represent quanta of energy and the squares represent atoms. The markers are moved among the squares randomly, one by one, and we ask for the distribution we should expect to find after a long time. That is, how many squares will be empty, how many will have one marker, and so on.

FIGURE 7.2 Shuffling on a checkerboard. Each square represents an atom and the checkers represent energy quanta. A square with 2 quanta simulates an atom in the second level of excitation. An empty square is an atom in the ground (unexcited) state.

And we ask to what extent the starting configuration will affect the results. If, for example, most of the markers are placed on one square at the outset, how long will that information be preserved by the system?

Suppose the squares are numbered $(1, 2, ..., N)$ and the moves consist of throwing dice to pick the donor square and throwing the dice again to pick the receptor square. Markers are moved one at a time. After each move, $n(q, t)$ is the number of squares with q markers, so q ranges from zero to Q. We refer to $n(q, t)$ as the "instantaneous occupancy function," and a square with q markers is said to be in the qth state of excitation.

A number of questions may be asked. The most general question is: What will be the shape of the averaged occupancy function, $n_q = \langle n(q, t) \rangle$, after the game has proceeded a large number of moves? More specifically, what is the most common number of chips we expect to find on each square? Another type of question is whether the average behavior depends in an essential way on the initial values, or on the rule by which the chips are selected for exchange. For example, suppose a random number of markers, from 1 to the total number on the donor square, are moved each time instead of taking just one chip from the donor square. Will this affect the average occupancy function?

Exercise 7.1

Draw a histogram of your prediction for the time-averaged histogram n_q of the Boltzmann process for a board with 4 squares and 8 quanta.

Then carry out a simulation with the chips on an array of squares. Find the histogram representing $n(q, t)$, the number of squares with q chips at each instant, and evaluate its time-average distribution, n_q.

One way to perform the simulation is as follows. Establish an array of N integers, $q(i)$, $(i = 1, ..., N)$, representing the number of chips on each square. Initialize the array in an arbitrary fashion so the sum of the integers equals the total number of chips, $\sum q(i) = Q$. Each step of the shuffling process consists of selecting i randomly; if $q(i) > 0$, take away a chip: $q(i) \to q(i) - 1$. Then randomly select the receptor square, j, and set $q(j) \to q(j) + 1$. The occupancy $n(q, t)$ is the number of squares with q chips at each time.

Before reading further, try to analyze this process and explain why the average histogram, n_q, has the shape it does. *Hint:* This problem can be made to resemble the coin-flipping run-length problem discussed earlier.

7.3 ANALYSIS OF THE BOLTZMANN PROCESS FOR A VERY LARGE SYSTEM

(a) Restatement of the Problem

There are many ways to analyze the shuffling of chips on the checkerboard in Figure 7.2. The following method is similar to the one we used in finding the distribution of run-lengths in a series of coin flips. It is an approximation that is valid for large numbers and is simpler to compute than the more exact method to be described later in this chapter.

The checkerboard is usually described as a square array of cells, each of which is defined by two indices, but we can equally well think of it as a string of squares, numbered from 1 to N. Suppose there are Q chips distributed among the squares. The mean number of chips on a square is $\langle q \rangle = Q/N$, and this is equivalent to the mean energy if we consider the squares to represent atoms and the chips to represent quanta of energy.

The occupancy function, $n(q, t)$, describes the number of squares with q chips at time t, and this can be visualized as a fluctuating histogram. The total number of chips is conserved; that is,

$$\sum_q qn(q, t) = Q. \tag{7.2}$$

We are interested in the time-average n_q, and we do not care *which* squares have q chips on the average, merely *how many* squares. (See Figures 7.3 and 7.4.)

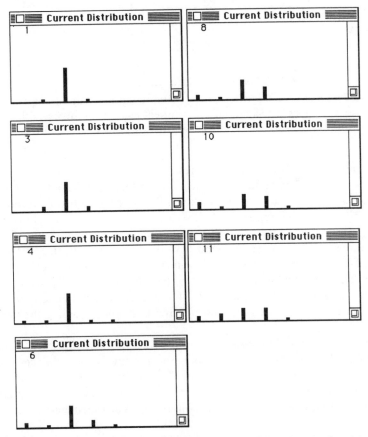

FIGURE 7.3 Sample evolution of the distribution of chips during the first 11 random transitions on a checkerboard with 16 squares. There are 32 chips, initially 2 on each square of the board. After the first transition, there were 14 squares with 2 chips, 1 with 3 chips, and 1 with 1 chip. An empty square had appeared after only 4 transitions.

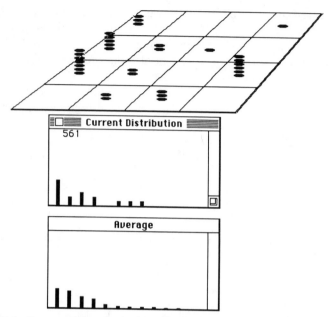

FIGURE 7.4 Status of the checkerboard with 16 squares and 32 chips after 561 transitions. The upper histogram shows the current numbers of squares with $0, 1, 2, \ldots$ chips and the lower histogram is the average over the history of the 561 transitions.

We now establish a different representation of the system, to make its similarity to coin flipping more obvious. Rather than specifying the number of chips on each square, we describe a state of the system by a list of symbols that contains the same information. The list contains just two types of symbol, S and C, representing squares and chips. An example is shown in Figure 7.5. The list contains N of the S symbols and Q of the C symbols. Each S is followed by a set of Cs corresponding to the number of chips on the corresponding square. The first symbol must be an S, but the remaining symbols can be arranged at random, and there are $N - 1 + Q$ in all. The Boltzmann process is equivalent to shuffling the positions of S and C symbols, holding the first one fixed.

Suppose we carry out a large number of exchanges and after each exchange we append the new list of symbols to a list of all previous exchanges. In this way, we gradually build a very long list. The analysis of this list is similar to the problem of determining the expected number of runs of various lengths in flipping a biased coin. We need only consider the correspondence

$$S \to \text{head}, \quad C \to \text{tail},$$

and imagine the list to represent the outcome of repeatedly flipping a die with $N - 1$ faces marked "S" and Q faces marked "C."

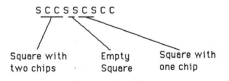

FIGURE 7.5 Algorithm for analyzing the Boltzmann process with a string of symbols representing squares (S) and chips (C). A run of two chips represents an atom with 2 quanta. The distribution of run-lengths gives the probabilities of excitation to various levels.

(b) Approximate Solution for Large Numbers

We can now treat the shuffling process as a run-length problem. We know that at any particular position in the list we have the mean probability

$$p(C) = \frac{Q}{N - 1 + Q} \tag{7.3}$$

of finding the symbol for a chip. [Also $p(S) = 1 - p(C)$ because each symbol must be either C or S.] If $Q \gg 1$ or $N \gg 1$, we can ignore the unity in the denominator. Dividing through by N, the fraction becomes

$$p(C) = \frac{\langle q \rangle}{1 + \langle q \rangle}, \tag{7.4}$$

where we have already defined $\langle q \rangle \equiv Q/N$ as the average number of chips per square, or the mean occupancy.

Now let us consider any symbol S in the list. The probability that it will be followed by q symbols of type C and then a symbol of type S is just $p(C)^q\{1 - p(C)\}$. This is the probability of finding q chips on that square, so the expected number of squares with q chips is

$$n_q = Np(C)^q\{1 - p(C)\}$$
$$= \frac{N\langle q \rangle^q}{(1 + \langle q \rangle)^{q+1}}. \tag{7.5}$$

Values of n_q for various $\langle q \rangle$ are shown in Table 7.1 for $N = 100$.

Exercise 7.2
Supply the missing steps of the preceding derivation and then verify Table 7.1.

TABLE 7.1   Expected Occupancies of Excited States, n_q

$\langle q \rangle$	$q = 0$	1	2	3	4	5
0.2	88.3	13.9	2.3	0.4	0.06	0.01
1.0	50	25	12.5	6.25	3.1	1.6
5.0	17.7	14.7	12.3	10.2	8.5	7.1

The expected number of empty squares (or atoms in the zeroth state) is

$$n_0 = \frac{N}{1 + \langle q \rangle}. \tag{7.6}$$

The ratio n_{n+1}/n_q, the relative occupancy of squares with $q+1$ and q chips is just $p(C)$, so we have

$$\frac{n_{q+1}}{n_q} = \frac{\langle q \rangle}{1 + \langle q \rangle}. \tag{7.7}$$

This ratio is a constant and it is less than unity, so the occupancies comprise a decreasing geometric series, or an exponential function. Thus we come to the remarkable result that the empty squares are the most numerous, regardless of the total number of chips on the board. When $N = Q$ and $\langle q \rangle = 1$, we expect that one-half the squares will be empty, on the average. One-quarter of the squares will be in the first state of excitation.

Exercise 7.3
Verify by simulation that the empty squares are the most numerous, on the average, after the system has hand time to relax and "forget" its starting configuration. This is true even when the average number of chips per square is greater than one.

When there are few quanta, and $\langle q \rangle$ is small, the drop-off is more rapid, and the number of squares with many chips becomes relatively smaller.

Exercise 7.4
Show that the solution for the occupancy ratios n_{q+1}/n_q given above leads to the proper value for the total number of quanta, in the limit as N and Q become large. That is,

$$\sum_{q=1}^{Q} q n_q = Q.$$

Exercise 7.5
Repeat the simulation of the Boltzmann process using different rules for the exchange of chips among squares. For example:

(a) A randomly chosen fraction of the chips is selected at each step. Show that this change of rule has no appreciable effect on the long-term average of the distribution function, but that it will lead to the more rapid relaxation of the system from its initial configuration.

(b) The receptor square is always adjacent to the donor square, in a randomly selected direction. Does this affect the result?

7.4 SHUFFLING KINETIC ENERGY AMONG ATOMS

The geometric-series behavior of n_q (which strictly holds only when the numbers of particles and quanta are very great) is a very general result. It is known as the Boltzmann distribution, and it does not depend on a particular choice of shuffling procedure.

As an illustration, let us focus on a gas of simple atoms that collide elastically with each other. We will suppose the sum of the energies of the colliding particles, E_1 and E_2, is divided randomly between the two particles. Their final energies (indicated by primes) are given by

$$E_1' = x(E_1 + E_2), \qquad E_2' = (1 - x)(E_1 + E_2), \qquad (7.8)$$

where x $(0 < x < 1)$ is a random variable representing the fraction given to the first atom.

The simulation may proceed as follows. Establish an array of cells representing the individual atoms. Give each atom a quantity of energy, represented by a number in the cell. Then choose a pair of cells at random, representing the colliding pair of atoms. Divide the total energy of the pair randomly between the two atoms and update the cell contents. Repeat the process many times and evaluate the average distribution of energy. The resulting average function n_E is independent of the shape of the distribution of x.

Exercise 7.6
Carry out the simulation using several distinct rules for selecting x and dividing the energy between the colliding pairs. Compare the resulting time-averaged distributions of kinetic energies n_E. *Hint:* See Appendix 2 for a discussion of drawing random numbers from a distribution.

The distribution n_E shows the relative occupancies of energy cells in a small range about the value E. In order to compute $N(E)$, the number of atoms with energy E, we must know how many such cells lie in each energy range. If we let $g_E dE$ be the number of cells in energy range $E-E + dE$, then we may evaluate $N(E)$ from the product of the occupancy and the number of cells in each energy range:

$$N(E)\, dE = n_E g_E\, dE. \qquad (7.9)$$

The method of counting the cells in each energy range depends on the nature of the gas and the answers to such questions as: Is the gas dense or dilute? Are the particles subject to the quantum exclusion principle? How many degrees of freedom do they have? We do not go deeply into this important problem but confine ourselves to a gas of N simple atoms in a spatial volume V. The kinetic energy of each atom is related to the momentum and mass by $E = p^2/(2m)$. The shell of momentum space corresponding to the energy range $E-E + dE$ has a volume $4\pi p^2\, dp$, and according to the Heisenberg uncertainty principle this shell is divided into cells of volume $\tau = Nh^3/V$. Thus the number of cells in the energy range $E-E + dE$ is given by $g_E\, dE \equiv 4\pi p^2\, dp/\tau$. Using the relation between E and p we can rewrite this as $g_E\, dE \sim \sqrt{E}\, dE$. So equation (7.9) gives $N(E) \sim n_E \sqrt{E}$, which is the product of a decreasing exponential and an

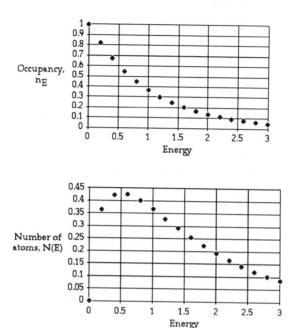

FIGURE 7.6 Occupancy n_E (upper diagram) and energy distribution $N(E)$ corresponding to Boltzmann exponential distribution of energies among randomly colliding atoms. The functions in the upper and lower diagrams are related by equation (7.9).

increasing function of E. the shapes of n_E and $N(E)$ are shown schematically in Figure 7.6.

7.5 HEIGHTS OF ATOMS IN A GRAVITATIONAL FIELD

As another application, consider a gas of atoms near the surface of a planet, where the gravity may be considered constant. Each atom has a potential energy $U = gmh$, where g is the acceleration of gravity, m is the atom's mass, and h is its height above the ground. If the total energy of all atoms is fixed, and the atoms are permitted to wander up and down colliding with each other, we expect the numbers of atoms with different potential energies, n_U to obey the Boltzmann distribution. Cells of higher energy (which are geometrically higher) will be less populated, and the atoms will be crowded toward the ground according to

$$n_U \sim \exp(-U) = \exp(-gmh). \qquad (7.10)$$

The following exercise is intended to verify the approximately exponential distribution of the particles.

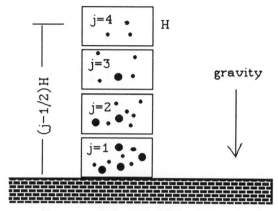

FIGURE 7.7   Simulation of a plane atmosphere by particles in boxes subjected to gravity. Particles of different mass are indicated with different sizes. See Exercise 7.7.

Exercise 7.7
Imagine a rigid stack of boxes, each of height H, in a gravitational field g. The midpoint of each box is at a height $(j - 1/2)H$, $(j = 1, 2, \ldots, J)$. Distributed among the boxes there are N particles of mass m. The gravitational energy of each particle is computed as though the particle were at the midpoint of the box, so $U = gm(j - 1/2)H$. The total energy of the system is the sum of the gravitational energies of all the particles.

Paticles are selected randomly and may move to another box, but the energy gained (if it moves down) or lost must be balanced by the displacement of one or more other particles, so that the total energy remains constant.

Starting from an arbitrary distribution, evaluate the total energy, then shuffle the particles repeatedly and compare the average height distribution of the particles with an exponential.

7.6   THE EXPECTED DISTRIBUTION FOR A FINITE SYSTEM

Thus far, our theory has been derived for infinite systems, while the preceding simulations have dealt with small or moderately large systems. Finite systems do not obey the exponential distribution expected for infinite systems, and we need a theory for finite systems. As is often the case in statistical physics, the algebra of the finite system is somewhat more complicated than for the infinite system.

We return to the checkerboard, with the remark that this is a model for a wide variety of physical phenomena, and our approach to understanding the shuffling process will provide insights in many areas.

For a specified number of squares (atoms) and chips (energy quanta), some distributions can be achieved in more ways than others. These are the distributions that will be observed more often in a simulation. The simplest

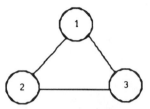

FIGURE 7.8 Schematic configuration graph for a system of three squares (atoms) and a single chip (energy quantum): $N = 3$, $Q = 1$. Each configuration corresponds to a different way of arranging the chips: In this case the chip is on a different square in each configuration. In all these configurations, there are 2 empty squares and 1 square with 1 chip. Hence the population distribution function is $n_0 = 2, n_1 = 1$ for all configurations. The lines indicate that the system can make the indicated transitions among configurations by the exchange of a single quantum.

interesting case has $N = 3$ squares. Suppose there is just 1 quantum, $Q = 1$. This quantum will be found on any one of three squares with equal probability.

We call each way of arranging the quanta a "configuration." There are three possible configurations, and we label them 1, 2, and 3 and represent them schematically on a triangular graph, as in Figure 7.8. The number inside each circle is the numerical label of the configuration. Each configuration corresponds to the same energy, and we assume that the system performs a random walk among the configurations and that all configurations are equally likely. If this is the case, then if we observe the system at an arbitrary time, we have an equal chance of finding it in any one of the configurations. For the simple triangular graph, this property seems obvious. Its applicability to more complex systems is known as the "ergodic hypothesis." This term (derived from the Greek words for *work* and *way*) was first used by Ludwig Boltzmann in 1887. Part of our task in this chapter is to illustrate the validity of this useful hypothesis for various systems.

As a more interesting example, suppose there are 2 chips to be distributed among the 3 squares ($N = 3$, $Q = 2$). (We do not distinguish between the chips, since they are assumed to be equivalent.) The six possible arrangements of the chips are shown in Table 7.2. Each entry is the number of chips on the

TABLE 7.2 Configuration of 2 Chips on 3 Squares

Configuration Number	Square		
	1	2	3
1	0	0	2
2	0	1	1
3	0	2	0
4	1	0	1
5	1	1	0
6	2	0	0

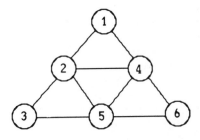

FIGURE 7.9 Configuration graph for a system of 3 squares and 2 chips $(N = 3, \ Q = 2)$. The configurations are listed in Table 7.2, and this graph indicates the possible transitions. For example, the system may move from configuration 1 to 2 in a single jump, but two jumps are required to go from 1 to 5.

corresponding square or, equivalently, the energy content of the atom. These configurations may be represented on a graph as in Figure 7.9. The lines indicate possible transitions involving a single quantum. We assume that these six configurations are equally likely and the system is shuffled randomly among them.

The population distributions for each configuration are shown in Table 7.3, where n_q is the number of atoms with q quanta. Note the configurations are of just two types: $(2, 0, 1)$ and $(1, 2, 0)$. We assume each configuration to be equally likely, so the average state of the system will correspond to the weighted average, or the straight sum, of all the distributions. These sums are indicated in the bottom row and are plotted in Figure 7.10. The population distribution is linear, with the empty square having the highest population. Thus we should expect atoms in the ground state to be the most frequent.

Exercise 7.8
Carry out a simulation with $N = 3$, $Q = 2$ and verify the expected frequencies implied by Table 7.3. (See program listing below for hints.)

As another example, we consider a 3-square system with $Q = 3$ quanta. Ten configurations are available. They are listed in Table 7.4 and they are graphed in Figure 7.11.

TABLE 7.3   Distributions for Each Configuration
$(N = 3, \ Q = 2)$

Configuration	n_0	n_1	n_2
1	2	0	1
2	1	2	0
3	2	0	1
4	1	2	0
5	1	2	0
6	2	0	1
Sum	9	6	3

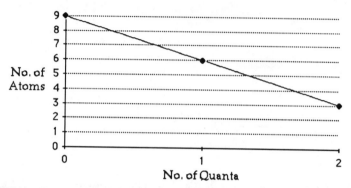

FIGURE 7.10   Population distribution for randomized system with $N = 3$, $Q = 2$ derived by summing over all configurations on the assumption that each configuration is equally likely to occur. The ratios of numbers of atoms in ground state, first level, and second level are 9:6:3.

TABLE 7.4   Configuration of 3 Chips on 3 Squares

Configuration Number	Square		
	1	2	3
1	0	0	3
2	0	1	2
3	0	2	1
4	0	3	0
5	1	0	2
6	1	1	1
7	1	2	0
8	2	0	1
9	2	1	0
10	3	0	0

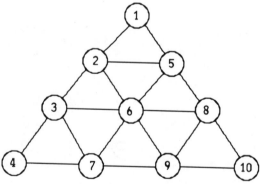

FIGURE 7.11   Configuration map for $N = 3$, $Q = 3$ showing allowed transitions.

TABLE 7.5   **Distributions for Each Configuration**
$(N = 3, Q = 3)$

Configuration	n_0	n_1	n_2	n_3
1	2	0	0	1
2	1	1	1	0
3	1	1	1	0
4	2	0	0	1
5	1	1	1	0
6	0	3	0	0
7	1	1	1	0
8	1	1	1	0
9	1	1	1	0
10	2	0	0	1
Sum	12	9	6	3

The population distributions for the possible configurations are listed in Table 7.5, along with the average population derived by summing with equal weights. As before, the average distribution is linear, with the ground state expected to be the most highly populated.

7.7   COMPUTER SIMULATIONS OF A SMALL SYSTEM

In this section, we illustrate two methods of simulating the history of a system consisting of three atoms and three energy quanta, $N = 3$, $Q = 3$. One method consists of tracking the system as it wanders through its configuration graph (Figure 7.11) in an asynchronous random walk. (See Chapter 5 for the definition of this mode of random walk.) The other method consists of randomly selecting a sequence of donor and receptor atoms and shuffling energy quanta from one set to the other.

Our purpose is to show that these methods give the same average behavior and to illustrate that the asynchronous walk on a configuration map is consistent with the ergodic hypothesis, described in Section 7.6.

(a)   First Method: Random Walk on a Configuration Graph

We first carry out an asynchronous random walk on the configuration graph and keep track of the transitions. At each time-step the system either undergoes a null transition or it jumps from its current configuration to a neighbor selected in accord with prescribed transition probabilities.

The configuration graph has ten nodes with 2, 4, or 6 connecting lines representing possible transitions. At the start of each time-step, the system is said to be in configuration i ($i = 1, \ldots, 10$) and during the subsequent interval

TABLE 7.6   Transition Probabilities for First Simulation, p_{ij}
($N = 3$, $Q = 3$)

i	$j = 1$	2	3	4	5	6	7	8	9	10
1	**5/7**	1/7	0	0	1/7	0	0	0	0	0
2	1/7	**3/7**	1/7	0	1/7	1/7	0	0	0	0
3	0	1/7	**3/7**	1/7	0	1/7	1/7	0	0	0
4	0	0	1/7	**5/7**	0	0	1/7	0	0	0
5	1/7	1/7	0	0	**3/7**	1/7	0	1/7	0	0
6	0	1/7	1/7	0	1/7	**1/7**	1/7	1/7	1/7	0
7	0	0	1/7	1/7	0	1/7	**3/7**	0	1/7	0
8	0	0	0	0	1/7	1/7	0	**3/7**	1/7	1/7
9	0	0	0	0	0	1/7	1/7	1/7	**3/7**	1/7
10	0	0	0	0	0	0	0	1/7	1/7	**5/7**

it undergoes a transition $i \to j$ ($j = 1, \ldots, 10$) with probability p_{ij}. The array in Table 7.6 summarizes the adopted transition probabilities p_{ij}. Each row gives the fractional probability of moving along each path out of the ith node. (See Figure 5.10 for a reminder on setting up these probabilities.) The assignment of probabilities starts from the node with the greatest number of exit paths. This is configuration number 6, which has 7 paths including the virtual path for the null transition. Each path is said to contribute the fraction $p = 1/7$ of a transition per time-step. All other paths to neighbors, $i \neq j$, were assumed to contribute the same amount to the probability of exit. Diagonal terms, $i = j$, correspond to null transitions and were evaluated so the terms across all rows would sum to unity. (This treatment is not fully satisfactory for small numbers of neighbors, but it is convenient and sufficiently accurate.)

During the walk, the computer program updates several arrays at each time-step. The elements of the array, $V[i]$ ($i = 1, \ldots, 10$), are initially set to 0, and they hold the number of times the system was observed in configuration i. After each time-step, the array is incremented to indicate that the system is in that node: $V[i] \to V[i] + 1$. The statistical properties of the system are contained in the list of numbers, $V[i]$, because they tell us the fractional time spent in each configuration. From this list we may compute n_q, the numbers of atoms with various amounts of energy.

While carrying out the simulation on this isolated system, we can also get the total number of transitions $i \to j$, which we designate as P_{ij}. (This is what we have called the "traffic" along the path $i \to j$). The results are shown in Tables 7.7–7.9 for a simulation consisting of 10,000 transitions. The first of these tables gives the counted values of the transition rates, P_{ij}. If we scan this array of transition rates, we note an important relationship that is approximately satisfied by the equilibrium system, namely, $P_{ij} = P_{ji}$. That is, the array is symmetric and each process goes approximately at the rate of its inverse. This is an illustration of the principle of "detailed balance in equilibrium."

TABLE 7.7   Transition Rates for First Simulation, P_{ij}
($N = 3$, $Q = 3$)

i	$j = 1$	2	3	4	5	6	7	8	9	10
1	**627**	143	0	0	141	0	0	0	0	0
2	153	**484**	140	0	136	153	0	0	0	0
3	0	143	**451**	124	0	148	143	0	0	0
4	0	0	130	**632**	0	0	126	0	0	0
5	131	150	0	0	**470**	143	0	133	0	0
6	0	146	150	0	145	**175**	144	152	143	0
7	0	0	139	132	0	139	**461**	0	140	0
8	0	0	0	0	135	151	0	**420**	151	139
9	0	0	0	0	0	146	137	141	**418**	152
10	0	0	0	0	0	0	0	150	142	**751**

TABLE 7.8   Mean Configuration Populations for First Simulation, $V[i]$
($N = 3$, $Q = 3$)

Configuration	1	2	3	4	5	6	7	8	9	10
Population	911	1066	1010	888	1027	1055	1011	996	994	1042

We also note that the sum of each vertical column and of each horizontal row is nearly equal to $1/10$ the total number of transitions. This indicates that the mean numbers of entries and exits from each configuration are nearly balanced. This is not the same as the detailed balance condition (which applies to every process, not merely every configuration) although it is one of the consequences of detailed balance.

Table 7.8 gives the mean populations of the configurations, and they range about the expected value, 1000. The measured root-mean-square deviation about 1000 is 55, which is somewhat larger than the expected value, $\sqrt{1000}$.

Exercise 7.9
Verify that the counted values of P_{ij} shown in Table 7.7 approximately equal the products of the probabilities in Table 7.6 and the mean number of visits to each configuration in Table 7.8; that is $P_{ij} = p_{ij}V[i]$. This relationship becomes exact in the limit of long simulations.

Finally, if we average the configurations (Table 7.5) using these population as weighting factors and compute the number of atoms with 0, 1, 2 and 3 quanta of energy, we find the results in Table 7.9, and they agree tolerably well with the theoretical numbers.

A program listing for carrying out this configuration walk with *HyperTalk* is shown below. The program is rather slow: It required about 15 minutes to do 10,000 steps on a Macintosh IIx, because it writes to the screen after every step, but it is very easy to set up and to modify. It could be made faster by eliminating the screen refreshing of the text fields.

TABLE 7.9 Computation of Mean Distribution for First Simulation, n_q
($N = 3$, $Q = 3$)

Configuration	Number of Excited Atoms				$V[i]$
	n_0	n_1	n_2	n_3	
1	2	0	0	1	911
2	1	1	1	0	1,066
3	1	1	1	0	1,010
4	2	0	0	1	888
5	1	1	1	0	1,027
6	0	3	0	0	1,055
7	1	1	1	0	1,011
8	1	1	1	0	996
9	1	1	1	0	994
10	2	0	0	1	1,042
Mean	11,786	9,269	6,104	2,941	
Expected	12,000	9,000	6,000	3,000	

HyperTalk Program Listing for the First Simulation

```
on mouseUp
-- uses variables oldconfig and newconfig to keep track of walk
-- stores counts in card fields
put item 1 of line 1 of cd fld "config" into oldconfig
-- initialize by reading first value
repeat with i = 1 to 10000   -- start loop for 10000 walks
        put getconfig (oldconfig,c2) into newconfig
-- find next configuration with function
-- listed below
        put newconfig into item 2 of line 1 of cd fld "config"
-- store in text field
-- now update visits count
        put 1+line  newconfig of cd fld "visits" into line  newconfig
of cd fld "visits"
-- update transition rate count
        get item newconfig of line  oldconfig  of cd fld "transition counts"
        put it+1 into item newconfig of line  oldconfig  of
cd fld "transition counts"
        put i into cd fld  "count"  – display running count
End repeat
End mouseUp
```

```
function getConfig c1, c2
-- user-defined to find next configuration to be visited
put the random of (70) into randT
-- draw random number in range 1 to 70
repeat with T=1 to 10 -- scan through cumulative probability table
      if randt<= item t of line c1 of cd fld "probabilities" then
            put t into c2  -- find value of new configuration
            exit repeat
      End if
End repeat
return c2  -- return value
End getconfig
```

(b) Second Method: Randomly Selecting Donor and Receptor Atoms

This method of simulation differs from the preceding because it selects a pair of atoms at random. Rather than imposing transition probabilites between pairs of configurations, we now assume that all atoms are equally likely to give and to receive a quantum of energy. (If these two simulation methods give similar results, we can conclude that the configuration transition probabilites are implicit in this assumption and in our method of counting configurations.)

The listing below (which excludes the graphics) shows the alternative computation. The program was designed to handle one specific system of atoms and quanta: $(N = 3, Q = 3)$. **Initialize** clears the arrays and inserts the desired amount of energy into the system. The energy contents of the atoms are held in an array **atom[i]** $(1 \leq i \leq N)$. The arrays **donorArray[i]** $(1 \leq i \leq N)$ and **receptorArray[i]** $(1 \leq i \leq N)$ tally the number of times each atom was selected to be a donor or a receptor. (This was done to verify that the selection process was a fair one.) The array **transition[i, j]** keeps track of the number of jumps from configuration i to configuration j, and gives P_{ij}. Array **meanConDist** keeps track of N_i, the number of visits to each configuration, and **meanDist** is the histogram of n_q, the mean numbers of atoms with different numbers of quanta.

The procedure **DoSwap** is the heart of the program. It records the previous configuration of the system, then randomly selects donor and receptor atoms. (They may be the same.) One quantum is then taken from the donor atom (if it has one to give) and is transferred to the receptor. The ID-number of the new configuration (see Table 7.3) is then found and the arrays are updated. Note the updating takes place even when the transition was a null transition, that is, if the donor has no energy to give or if the donor and receptor were the same atom.

The main part of the program (at the end) consists of seeding the random number with a number that will be different each time the program is started, and then a call to the procedure **Initialize** and then a loop that calls procedure **DoSwap** many times. Finally, the results are displayed.

TABLE 7.10   Counted Transition Rates for Second Simulation ($i \rightarrow j$)
($N = 3, Q = 3$)

$i \backslash j$	1	2	3	4	5	6	7	8	9	10
1	**780**	118	0	0	113	0	0	0	0	0
2	111	**558**	105	0	119	114	0	0	0	0
3	0	114	**443**	104	0	106	99	0	0	0
4	0	0	113	**791**	0	0	110	0	0	0
5	120	107	0	0	**570**	114	0	96	0	0
6	0	110	113	0	109	**315**	101	106	116	0
7	0	0	92	119	0	102	**597**	0	113	0
8	0	0	0	0	95	104	0	**567**	119	110
9	0	0	0	0	0	115	116	105	**560**	125
10	0	0	0	0	0	0	0	123	112	**850**

TABLE 7.11   Mean Distribution for Section Simulation
($N = 3, Q = 3$)

	n_0	n_1	n_2	n_3
Simulation	12,140	8,830	5,920	3,110
Expected (Table 7.5)	12,000	9,000	6,000	3,000

In a run of 10,000 swaps, the three atoms acted as donors in 3247, 3451, and 3302 cases, respectively, while they acted as receptors in 3306, 3387, and 3307 cases. Thus the random selection process seems to have worked fairly. The populations of the 10 configurations were all nearly equal to the expected value, 1000, consistent with the assumption that the system executed a random walk through the configuration graph. The counted transition rates, P_{ij}, are listed in Table 7.10, giving the number of transitions $i \rightarrow j$. Finally, the weighted populations, n_q, were nearly equal to the expected values, as shown in Table 7.11.

Program Listing for the Second Simulation

Program SmBoltz; {Simulation of Boltzmann process for small board}
{with tracking of transitions}

```
        const
                MaxQ = 3;
                N = 3;  {number of atoms or squares}
        var
                MeanDist : array[0..maxQ] of integer;
{population distributions}
                atom, donorarray, receptorarray : array[1..N] of integer;
{atoms holding quanta}
                transition : array[1..10, 1..10] of integer;
{configuration transitions}
```

```
                MeanConDist : array[1..10] of integer;
{keep track of configurations}
                oldCon, newCon, step, donor, receptor : integer;
                draw_rect : rect;  {for display of results}
                q : integer;  {number of quanta}


Procedure initialize;  {clear arrays and inject initial quanta into system}
        var
                i, j : integer;
        Begin
        setRect(Draw_rect, 5, 35, 510, 340);
        setDrawingRect(draw_rect);
        showdrawing;
        for i := 1 to N do
                Begin
                atom[i] := 0;
                donorarray[i] := 0;
                receptorarray[i] := 0;
                End;
        for i := 0 to maxQ do
                Begin
                nowdist[i] := 0;
                meandist[i] := 0;
                End;
        for i := 1 to 10 do  meancondist[i] := 0;
        for i := 1 to 10 do
                for j := 1 to 10 do  transition[i, j] := 0;
        atom[1] := 2;  {load quanta}
        atom[2] := 1;
        q := 3;
        End;


Procedure updateHists;  {after each transition, do the bookkeeping}
        var
                i : integer;
        Begin {increment counters}
        donorArray[donor] := donorArray[donor] + 1;
        receptorArray[receptor] := receptorArray[receptor] + 1;
        transition[oldcon, newcon] := transition[oldcon, newcon] + 1;
{increment transition array}
        meanConDist[newcon] := meancondist[newcon] + 1;
{increment tally of configurations occupied}
        for i := 1 to N do  meanDist[atom[i]] := meanDist[atom[i]] + 1;
{update energy distribution}
        End;
```

```
Function findConfig : integer;
{from occupation numbers, find the configuration  ID number}
        Begin
        if atom[1] = 3 then findconfig := 10
        else if atom[1] = 2 then
                if atom[2] = 1 then findConfig := 9
                else  findConfig := 8
        else if atom[1] = 1 then
                 if atom[2] = 2 then findConfig := 7
                else if atom[2] = 1 then  findConfig := 6
                else findConfig := 5
        else if atom[1] = 0 then
                if atom[2] = 3 then findConfig := 4
                else if atom[2] = 2 then findconfig := 3
                else if atom[2] = 1 then findconfig := 2
                else findConfig := 1;
        End;


Procedure DoSwap; {carry out a selection of donor and receptor}
        Begin
        oldcon := newcon;  {set previous new to old}
        donor := trunc(N * (abs(random) / 32767)) + 1;
{select integer from 1 to N,  number of atoms}
        receptor := trunc(N * (abs(random) / 32767)) + 1;  {ditto}
        if donor > N then donor := N;  {error trap}
        if receptor > N then receptor := N;  {ditto}
        if atom[donor] > 0 then   { atom has some energy to give}
                Begin
                atom[donor] := atom[donor] - 1;
                atom[receptor] := atom[receptor] + 1;
                End;
        newcon := findConfig; {get ID number of the new configuration}
        if step > 10 then updateHists; {skip the first 10 transitions}
{Note, we then update the histogram in every call to DoSwap; }
{The null transitions (i.e. donor = empty or donor = receptor) are counted}
{in the averaging process}
        End; {doSwap}


Procedure displayMeans; {graphical output}
        var
                i, j : integer;
        Begin
{This will depend on the user and hardware.}
        End;
```

```
Begin {main SmBoltz}
        randseed := tickcount;
        hideall;
        initialize;
        for step := 1 to 10010 do doswap; {set the number of transitions}
        displayMeans;
End.
```

(c) Comparison of the Simulations

The agreement between the two methods of simulating the random shuffling of quanta among atoms is quite good. It illustrates that the problem may be formulated in a variety of fashions. Once the graph has been set up and the question of setting up the transition probabilites has been settled, the random walk on a configuration graph (First Method) is much easier to program than the random selection of donor and receptor atoms (Second Method). This is evident from the shortness of the Hypertalk program that did the work. The use of configuration graphs is simpler because we only need to track one variable, namely, the index of the current and the previous configuration. The program is logically simpler because the conservation of energy is built into the configuration map, and we do not need to go through the process of seeing whether an atom has energy to give.

7.8 GENERAL EXPRESSION FOR THE NUMBER OF CONFIGURATIONS

In order to generalize our results, we need an expression for the number of configurations in an arbitrarily large system of N atoms and Q quanta. Let $\Omega(N, Q)$ be the total number of configurations consistent with prescribed values of N and Q. We have already found $\Omega(3, 1) = 3, \Omega(3, 2) = 6, \Omega(3, 3) = 10$.

(a) Exact Expression for $\Omega(N, Q)$

The value of $\Omega(N, Q)$ plays a key role in statistical physics, as we see in Chapter 8. In Exercise 7.10, the reader is to verify that Ω can be represented by Table 7.12, which has the appearance of Pascal's triangle laid on its side. Note the symmetry of the table, and verify that the entries obey the following sum rule:

$$\Omega(N, Q) = \sum_{Q'=0}^{Q} \Omega(N - 1, Q'). \tag{7.11}$$

Exercise 7.10
Following the method of counting configurations outlined earlier in this chapter, verify several entries in Table 7.12.

TABLE 7.12 Number of Configurations, $\Omega(N, Q)$ (N Atoms with Q Quanta)

Q	$N =$ 1	2	3	4	5	6	$\cdots$	12
0	1	1	1	1	1	1		1
1	1	2	3	4	5	6		12
2	1	3	6	10	15	21		78
3	1	4	10	20	35	56		364
4	1	5	15	35	70	126		1,365
5	1	6	21	56	126	252		4,328
6	1	7	28	84	210	462		12,376

Thus the number of configurations for a system with N atoms can be obtained by summing over lower energies the numbers corresponding to a system with $N - 1$ atoms. The explanation of this rule is straightforward. Consider a system with $N = 5$ and $Q = 2$. Now add one atom ($N \to 6$) without altering the total energy Q. This additional atom can accept any amount of energy, not exceeding Q. Consider the possibilities, illustrated in Figure 7.12. If the new atom accepts zero energy, the original 5 atoms retain 2 quanta, which may be arranged in $\Omega(5, 2) = 15$ ways. If it accepts 1 quantum of energy, the original atoms will share one unit, and this quantum can be arranged in $\Omega(5, 1) = 5$ configurations. If the new atom accepts 2 quanta, the original 5 are left without any quanta, giving $\Omega(5, 0) = 1$ configuration. Summing, the total number of configurations for the system of 6 atoms and 2 quanta is

$$\Omega(6, 2) = \Omega(5, 0) + \Omega(5, 1) + \Omega(5, 2),$$

and by induction we have the sum rule expressed above.

Exercise 7.11 verifies the identity of $\Omega(N, Q)$ to the entries of the Pascal triangle and hence to the binomial coefficients. This suggests, according to the results of Chapter 3, that we can view this counting of configurations as a

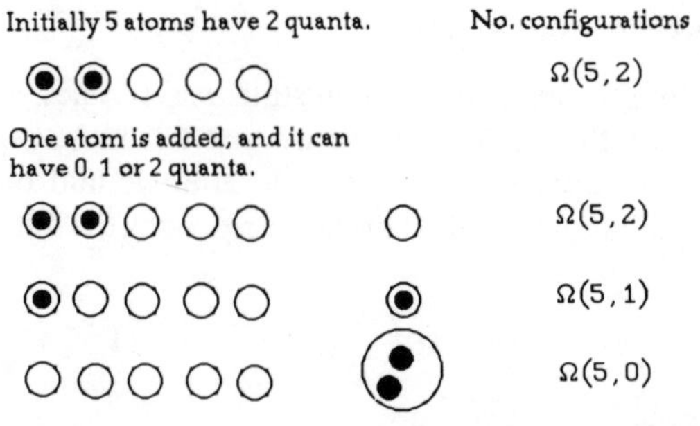

FIGURE 7.12 Schematic representation of system of 6 atoms and 2 quanta being built from a system of 5 atoms, for the purpose of computing total configurations: $\Omega(6, 2) = \Omega(5, 0) + \Omega(5, 1) + \Omega(5, 2)$.

problem of counting paths in a random walk. This analogy provides an explicit expression for Ω.

Exercise 7.11
By considering the difference of $\Omega(N, Q)$ and $\Omega(N - 1, Q)$, show that the sum rule for $\Omega(N, Q)$ leads to the rule defining the Pascal triangle, $\Omega(N, Q) = \Omega(N, Q - 1) + \Omega(N - 1, Q)$.

We start with a system of 1 atom and 0 quanta ($N = 1$, $Q = 0$), and we wish to build up a system of N atoms and Q quanta. The sequence in which we add atoms or quanta is arbitrary. Each sequence gives a particular configuration of the (N, Q) system. This building up of a sequence is similar to a random walk starting at location $(1, 0)$ and proceeding to (N, Q). Since $N - 1$ is the number of added particles and Q is the added number of quanta, we have, from our solution of the random walk in Chapter 4,

$$\Omega(N, Q) = \frac{(N - 1 + Q)!}{(N - 1)! Q!}, \tag{7.12}$$

and calculation verifies that this gives the entries of Table 7.12.

(b) Approximation for Large Systems

The factorials in the expression for $\Omega(N, Q)$ may be simplified using Stirling's approximation for the logarithm,

$$\ln M! = (M + 0.5) \ln M - M + 0.5 \ln 2\pi + O(1/M) \qquad (M \gg 1), \tag{7.13}$$

where $\ln M$ is the natural logarithm. Although the formula is valid in the limit of large M, Table 7.13 demonstrates that it works quite well for values as small as 8. (Compare the two bold faced columns.) In fact, we can use this abbreviated version with fair precision (as shown in the final column),

$$\ln M! \approx M \ln M - M \qquad (M \gg 1). \tag{7.14}$$

This greatly simplifies our later analysis.

TABLE 7.13 Test of Stirling's Approximation for ln M!

M	$M!$	$\ln M!$	$M \ln M$	$(M + 0.5)\ln M - M$	Eq. (7.13)	$M \ln M - M$
3	6	**1.79**	3.30	0.85	**1.42**	0.30
4	24	**3.18**	5.55	2.24	**2.81**	1.55
5	120	**4.79**	8.05	3.85	**4.42**	3.05
6	720	**6.58**	10.75	5.65	**6.22**	4.75
7	5,040	**8.53**	13.62	7.59	**8.17**	6.62
8	40,320	**10.60**	16.64	9.68	**10.25**	8.64

If we set $N - 1 \approx N$, the abbreviated version of Stirling's approximation gives the following useful expression for the logarithm of the number of configurations:

$$\ln \Omega \approx (N + Q)\ln(N + Q) - N \ln N - Q \ln Q. \qquad (7.15)$$

Exercise 7.12
Verify the derivation of (7.15) and test the expression numerically for $N = 10$, $Q = 5$ by comparing it with the exact value.

7.9   ERGODIC PROPERTY OF MIXED SYSTEMS

Not all systems are ergodic, and it is not a simple matter to prove ergodicity, but the assumption that a particular system (such as a collection of colliding atoms) is ergodic can be a very powerful device. If we can assume ergodicity for a system, we can compute its mean state by simply averaging over all the configurations available to the system. The meaning of the ergodic property may be described in two ways.

(a) Equivalence of Two Types of Average

In discussing the average of a particular property such as the energy per particle, we have not always distinguished between two possible ways of taking the average:

(1) The average of all particles at a give instant *versus*
(2) The time average for a single particle, followed over a long interval.

According to the ergodic hypothesis, which appears to be valid for most well-mixed physical systems, these two types of average give the same result. For example, if we follow a single square on the board at N different times and tabulate the number of times it has q quanta, we can derive from these data the distribution

$$n_q = \frac{1}{N} \sum_{i=0}^{N} n(q, t_i). \qquad (7.16)$$

This is essentially the probability that it will hold q quanta at an arbitrary time. On the other hand, we can examine a large number of squares at a given instant t_0 after the system has run long enough to become well mixed. If there are M squares, we can count the quanta on each square, q_i, and define the distribution:

$$n'_q = \frac{1}{M} \sum_{i=0}^{M} n(q_i, t_0). \qquad (7.17)$$

If the ergodic property holds, and if we have given the system sufficient time to forget the starting point and to become well mixed, these two distributions are, on average, the same; that is, $n_q = n'_q$. The following exercise illustrates this property.

Exercise 7.13
Return the simulation of Exercise 7.1 for a large system with, say, 100 atoms and 50 quanta. At time t_0, after the system has run long enough to become well mixed, record the instantaneous distribution n'_q, defined above. The follow a particular atom and record the time-averaged distribution n_q. Verify that if this is done often enough, the two types of distribution approach the same limiting shape.

(b) Uniform Filling of Configuration Space

We may also describe the ergodic property by referring to configuration graphs such as shown in Figure 7.7, 7.8, and 7.10. The simulation of the previous section demonstrated that if the walker is trapped in a finite graph and carries out an asynchronous walk, it will in the course of time visit all points of the lattice equally frequently. This is the ergodic property in a slightly different guise.

In succeeding chapters, we often call on the ergodic hypothesis.

REFERENCES

Baker 1986
Kittel and Kroemer 1980
Reichl 1980
Reif 1965

8

EQUILIBRIUM AMONG SYSTEMS

This chapter studies the approach to thermal equilibrium between small systems of atoms, and it introduces the concepts of temperature and entropy, pointing out the need for a distinction between large and small systems when discussing these thermodynamic concepts.

8.1 SIMULATION OF TWO SYSTEMS IN THERMAL CONTACT

(a) Specification of the Systems

For the sake of visualization, we represent an isolated system of atoms by a checkerboard with N_i squares (representing atoms) and $Q_i(0)$ equal quanta of energy, ε. An isolated system, D_i, has a fixed number of particles and will evolve with constant energy, so $Q_i(t) = Q_i(0)$. We define the instantaneous distribution, $n_i(q, t)$, as the number of squares with q chips at time t. The mean numbers are described by a time-averaged distribution, $\langle n_i(q, t) \rangle$, which is uniquely specified by N_i and $Q_i(0)$. We call this the equilibrium state of the system, noting that this name implies isolation and long-time averaging.

But how do we determine the system parameters quantitatively? Each atom is assumed to have a known mass, m so the total mass of the system is mN_i, and we can in principle determine N_i by measuring the total mass. Determining the energy content $Q_i(0)$ requires putting the system into thermal contact with a large reservoir, and we now turn our attention to thermal contact.

$$Q_1(t) + Q_2(t) = Q_s, \text{ a constant.}$$

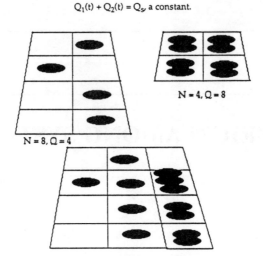

$N = 4, Q = 8$

$N = 8, Q = 4$

$N = 12, Q = 12$

FIGURE 8.1 Two systems (upper) in thermal contact that are combined (lower) to make a single system.

(b) Thermal Contact

Put this system ($i = 1$) in thermal contact with another ($i = 2$) and follow their history as they exchange energy (Figure 8.1). Let the total number of atoms in the new system be $N_s = N_1 + N_2$. At each step of a simulation, a donor square is chosen at random from either system and then a receptor square is chosen the same way. One quantum is then displaced from the donor to the receptor and the process is repeated. The energy quanta (chips) will migrate from one board to another, and $Q_1(t)$ and $Q_2(t)$ will vary with time. The sum (representing the total energy of the system of two boards) remains constant:

$$Q_1(t) + Q_2(t) = Q_s, \quad \text{a constant.}$$

Exercise 8.1

Carry out a simulation of the two-board system, with the following parameters: $N_1 = 8$, $N_2 = 4, Q_1(0) = 4, Q_2(0) = 8$. Use a method similar to that of the simulation in Exercise 7.1 and keep track of the chips on each of the boards. To choose the donor and receptor squares, set up a list of $N_s = N_1 + N_2 = 12$ squares and choose integers randomly from the range $1 < i < 12$. Start from a random assignment of chips to squares, $q(i)$, and compute time averages of the distributions $\langle n_1(q, t) \rangle$ and $\langle n_2(q, t) \rangle$, where the average is over all the preceding N time-steps.

Before reading further in the text, examine the results of your simulation, and seek to answer the following questions:

(a) What are the time-average ratios of the number of chips to number of squares for each subset, $\langle Q_1(t) \rangle / N_1$ and $\langle Q_2(t) \rangle / N_2$? How are they related to each other?

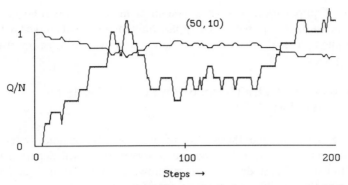

FIGURE 8.2   Simulation of a pair of subsystems in thermal contact. The larger system ($N = 50$) shows smaller fluctuations than the smaller system ($N = 10$). Because they are in thermal contact and can exchange energy but not mass, the energy per particle, Q_i/N_i, fluctuates about the same mean value in the two subsystems.

(b) What are the shapes of the time averages of the distributions in the two systems, $\langle n_1(q, t) \rangle$ and $\langle n_2(q, t) \rangle$, after a long time?

The answers to the questions in Exercise 8.1 concerning the time averages of the ratios $\langle Q_1(t) \rangle/N_1$ and $\langle Q_2(t) \rangle/N_2$ are not difficult to find if we do a "thought experiment." Imagine the two boards combined into a single board, as in Figure 8.1. The donors and receptors are chosen randomly, so there is nothing to distinguish one square from any other. This implies that, after a long simulation, each square ought to have the same mean number of chips, $\langle q \rangle = Q_s/N_s$. So the average amount of energy possessed by each subset should be proportional to the number of squares in that subset, or $\langle Q_i \rangle = \langle q \rangle N_i$.

Figure 8.2 shows the energy contents of two systems with 50 and 10 squares in simulated thermal contact, containing a total of 50 quanta of energy. The number of chips per square, $Q_i(t)/N_i$, is plotted for each board as a function of time, and we see that the curves approach each other and then fluctuate about the mean values of the combined systems, $\langle q \rangle = Q_s/N_s$.

Based on such experiments, we adopt the following axiom and corollary:

Axiom: *In a system consisting of any number of subsystems (boards) in contact with each other, all squares have the same mean number of chips, $\langle q \rangle$. (All atoms have the same mean energy.)*

Corollary: *If one of the subsystems starts with more than its fair share of chips, it will tend to give up chips to the rest of the system so that its average energy per atom will approach the system average.*

The tendency toward equal energy per atom in the two systems is a typical example of equipartition of energy among stochastically interacting systems. Each subsystem receives a share of energy proportional to the number of atoms it contains. Equipartition in such a system is merely a consequence of the shuffling rule we have imposed, and it applies only to the time-averaged behavior. At any moment, a small system may deviate a long way from equipartition.

8.2   ENERGY FLUCTUATIONS BETWEEN SYSTEMS IN THERMAL CONTACT

We analyze the simulated thermal contact of two Boltzmann systems and evaluate the relative probability that an observed state will depart from the equipartition state. These departures are known as "fluctuations," and we are able to derive the likelihood of fluctuations of various sizes.

(a) Equal Systems

First, consider two systems D_1 and D_2 defined by equal numbers of atoms, $N_1 = N_2 = 3$, and a total of $Q_1 + Q_2 = 6$ quanta, which they are free to exchange with each other. We examine the consequences of various assumed partitionings of these 6 quanta and show that $Q_1 = Q_2 = 3$ leads to the highest probability. The second and fourth columns of Table 8.1 show the number of possible configurations, $\Omega(3, Q_i)$, for each of the systems with various partitionings consistent with $Q_1 + Q_2 = 6$. The fifth column is the product $\Omega(3, Q_1)\Omega(3, Q_2)$ and represents the total number of configurations for the combined system of 6 atoms with a prescribed energy in each system. The sum of the entries of the fifth column is the total number of configurations available for $Q = 6$, $N = 6$ with an arbitrary partitioning of the energy. The final column shows the normalized probabilities, computed by dividing entries of the fifth column by their sum.

Exercise 8.2
Using the formula given earlier, verify that $\Omega(6, 6) = 462$, the sum of column 5 in Table 8.1.

TABLE 8.1   Composite Boltzmann System
$$(N_1 = N_2 = 3; \ Q_1 + Q_2 = 6)$$

Q_1	$\Omega(3, Q_1)$	Q_2	$\Omega(3, Q_2)$	$\Omega(3, Q_1)\Omega(3, Q_2)$	$p(Q_1, Q_2)$
6	28	0	1	28	0.061
5	21	1	3	63	0.136
4	15	2	6	90	0.195
3	10	3	10	100	0.216
2	6	4	15	90	0.195
1	3	5	21	63	0.136
0	1	6	28	28	0.061
				Sum = 462	Sum = 1.000

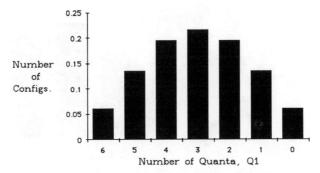

FIGURE 8.3a   The number of configurations depends on the amount of energy held by the first atom. Based on Table 8.1.

As the systems wander at random among the sample space of possible configurations, they are most often found in the state of equipartition (0.216 of the total time). At other times, one of the subsystems will have more than half the energy. For example, one or the other of the subsystems will have all the energy $2 \times 0.061 = 0.122$ of the time. Because the probabilities are symmetric about the maximum, each subsystem will have an excess energy just one-half of the time, as shown in Figure 8.3a.

Exercise 8.3
Following the example of Table 8.1, construct a table showing the numbers of configurations for a composite system consisting of two subsystems with $N_1 = 6$, $N_2 = 6$, and $Q_1 + Q_2 = 12$. This method can be used to build tables for very large systems.

(b) Unequal Systems

Now let us consider a pair of unequal subsystems with 2 and 4 particles, respectively. In this case, equipartition looks a little different. Table 8.2 and Figure 8.3b show the results in a format similar to Table 8.1.

TABLE 8.2   Composite Boltzmann System
$$(N_1 = 2, \; N_2 = 4; \; Q_1 + Q_2 = 4)$$

Q_1	$\Omega(2, Q_1)$	Q_2	$\Omega(4, Q_2)$	$\Omega(2, Q_1)\Omega(4, Q_2)$	$p(Q_1, Q_2)$
4	5	0	1	5	0.040
3	4	1	4	16	0.127
2	3	2	10	30	0.238
1	2	3	20	40	0.317
0	1	4	35	35	0.278
				Sum = 126	Sum = 1.000

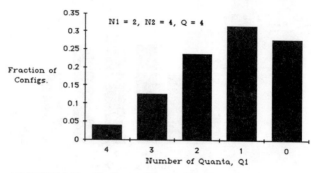

FIGURE 8.3b   Similar to Figure 8.3a, based on Table 8.2.

The subsystem with the greater number of atoms is expected to have more energy. The most likely configuration is one in which the first subsystem has only 1 quantum while the second has 3 quanta. Because the distributions are skewed, this is not the same as the average configuration. We may compute the averages as follows:

$$\langle Q_1 \rangle = \sum_{Q_1} Q_1 p(Q_1, Q_2) = 4(0.04) + 3(0.127) + 2(0.238) + 1(0.317) = 1.334,$$

$$\langle Q_2 \rangle = \sum_{Q_2} Q_2 p(Q_1, Q_2) = 4(0.278) + 3(0.317) + (0.238) + 1(0.127) = 2.666.$$

Thus the average energy per particle, $Q_i/N_i = 2/3$, is the same in the two subsystems, in accordance with the equipartition suggested by the earlier simulations. From this table we may also assess the probabilities of various departures from equilibrium.

(c) Single Atom in Contact with Large System

Now let us consider a single atom, D_1, in contact with a large system, D_2, consisting of N_2 atoms and a specified number of quanta, Q. We are interested in $p(Q_1)$, the probability that the atom will have Q_1 quanta, and we may proceed exactly as before. Table 8.3 shows the results.

According to Table 8.3, the single particle is most often (with $p = 0.668$) without any quanta, as this leads to the greatest number of configurations for the remaining part of the 13-particle system. Its likelihood of holding various numbers of quanta is approximately exponential, as Figure 8.3c and 8.3d illustrate. Its mean energy is just $6/13 = \sum Q / \sum N$, as the reader is asked to verify in Exercise 8.4.

Exercise 8.4
Using the numbers in Table 8.3, verify that $\langle q \rangle = \sum Q_1 p(Q_1) = 6/13$.

TABLE 8.3   Single Atom and Large Boltzmann System
$$(N_1 = 1, \ N_2 = 12; \ Q_1 + Q_2 = 6)$$

Q_1	$\Omega(1, Q_1)$	Q_2	$\Omega(12, Q_2)$	$\Omega(1, Q_1)\Omega(12, Q_2)$	$p(Q_1)$
6	1	0	1	1	0.000054
5	1	1	12	12	0.00065
4	1	2	78	78	0.0042
3	1	3	364	364	0.020
2	1	4	1,365	1,365	0.074
1	1	5	4,328	4,328	0.234
0	1	6	12,376	12,376	0.668
				Sum = 18,524	Sum = 1.000

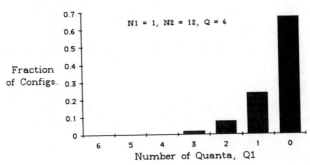

FIGURE 8.3c   Similar to Figure 8.3a, based on Table 8.3. (See also Figure 83d.)

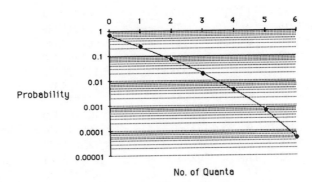

FIGURE 8.3d   Probability that a single atom will have various numbers of quanta. Based on Table 8.3.

Exercise 8.5

(a) Carry out a simulation of the thermal contact between two Boltzmann systems ($N_1 = 12$ and $N_2 = 1$, $Q_1 + Q_2 = 6$) and test the predictions of Table 8.3 and the approximately exponential probability that the single particle holds various amounts of energy.

(b) By changing the total energy, $Q_1 + Q_2$, investigate how the steepness of the drop-off depends on the total energy.

(d) Large Composite Boltzmann System

When the composite system contains a large number of particles and quanta the probabilities become sharply peaked around the highest value. As a consequence, if the system is observed at random times, its configuration will correspond closely to the most likely configuration. This tendency is illustrated in Table 8.4, for a system of 24 particles and 12 quanta (see Figure 8.4).

We see from Table 8.4 that the systems more than a few steps from the most likely system correspond to a small fraction of the total probability. Equipartition becomes more likely as the system grows. Putting this another way, in large systems the sum of the number of configurations, Ω, is primarily made up from the number of configurations corresponding to equipartition.

Exercise 8.6

Consider a system of 100 particles and 100 quanta to be a composite of two systems of 50 particles. Use the equation for $\Omega(N, Q)$ to show that the major contribution to $\Omega(100, 100)$ comes from the single product $\Omega(50, 50)\Omega(50, 50)$ corresponding to equipartition.

TABLE 8.4   Large Boltzmann System
$$(N_1 = 12, \, N_2 = 12; \, Q_1 + Q_2 = 12)$$

Q_1	$\Omega(12, Q_1)$	Q_2	$\Omega(12, Q_2)$	$\Omega(12, Q_1)\Omega(12, Q_2)$	$p(Q_1)$
12	1,352,078	0	1	1,352,078	0.0061
11	705,432	1	12	8,465,184	0.0101
10	352,716	2	78	27,511,848	0.0330
9	167,960	3	364	61,137,440	0.0733
8	75,582	4	1,365	103,169,430	0.1236
7	31,824	5	4,328	137,734,272	0.1651
6	12,376	6	12,376	153,165,376	0.1836
5	4,328	7	31,824	137,734,272	0.1651
4	1,365	8	75,582	103,169,430	0.1236
3	364	9	167,960	61,137,440	0.0733
2	78	10	352,716	27,511,848	0.0330
1	12	11	705,432	8,465,184	0.0101
0	1	12	1,352,078	1,352,078	0.0016
				Sum = 834,451,800	Sum = 1.000

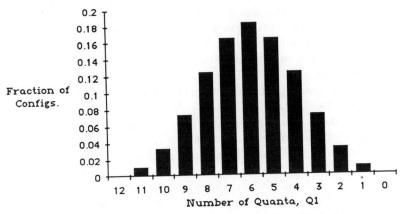

FIGURE 8.4   Similar to Figure 8.3a, based on Table 8.4.

Exercise 8.7

In a like manner, consider a system of 100 particles and 100 quanta to be a composite of two systems of 70 and 30 particles, respectively. Show that the major contribution to $\Omega(100, 100)$ comes from the single product $\Omega(70, 70)\Omega(30, 30)$ corresponding to equipartition.

To summarize, suppose we consider two system D_i and D_j, which we place in thermal contact so they combine to form the single system D_s. In the limit of large systems, the values of Q_i and Q_j will be partitioned between the subsystems in proportion to N_i and N_j, the numbers of particles in each subsystem. This is equipartition. For small systems, the probabilities are not so sharply peaked and the configuration will fluctuate about the equipartition state. The likelihood of each configuration will follow probabilities such as those shown in the preceding tables.

8.3   THE ZEROTH LAW OF THERMODYNAMICS

Our discussion has focused on the statistical properties of microscopic systems consisting of small numbers of particles, and we wish now to explore the application of these ideas to thermodynamics, which deals with measurable properties, such as temperature and pressure. The essential link between the two approaches is in the apparatus used to measure thermodynamic quantities. This apparatus (thermometers and springs) is macroscopic and consists of very large numbers of atoms (at least 10^{20}). Only if we can deal with quantities that can be measured with such apparatus can we speak of "thermodynamics."

The concept of temperature is central to thermodynamics, and we need to approach its definition from two directions: empirical and theoretical. For large systems, $N \gg 1$, we approach the problem empirically and may introduce a thermometer, whose mechanical state depends on temperature or, more specifically, the energy content per particle in the system. (The thermometer must be described in detail if this statement is to avoid circularity.) When we discuss small systems, in which fluctuations can be important, this empirical temperature becomes ill-defined and the concept must be used with some caution. We encounter the difficulty head-on when we attempt to measure the energy content of a small system. This difficulty will not prevent us from adopting a well-defined *theoretical* definition in a later section.

We have said that an isolated Boltzmann system is completely prescribed once the number of atoms and the amount of energy it contains are specified. But we cannot directly measure the energy content of a collection of atoms. The best we can do is put the system into thermal contact with a large system (reservoir) whose temperature has already been determined. We then assume that the temperature of the specimen will approach the temperature of the reservoir. But if the specimen is very small, its energy content will fluctuate during thermal contact with the reservoir. The actual value at the moment we pull the two systems apart will be a matter of chance. Thus, we cannot consider this empirical temperature to be a well-defined quantity for a small isolated system; we must retain thermal contact with the reservoir and adopt an average over a large number of small fluctuations.

Recognizing the limitations on precision that are imposed by fluctuations in small systems, let us look in more detail at this process.

(a) Rank-Ordering the Systems According to Temperature

We start with the sequence suggested by Ehrlich (1981). Consider a set of isolated systems, D_i, each characterized by fixed numbers of atoms, N_i, and an initial number of energy quanta, $Q_i(0)$. These systems can be isolated [so $Q_i(t)$ will remain constant] or they may be put into thermal contact, permitting energy exchange.

The principle of equipartition can provide us with a way to place them into a *temperature order*. Let us put two of the systems, D_1 and D_2, into thermal contact with each other, as in the two-board simulation in Exercise 8.1. Each system retains its original atoms (squares), but it may exchange energy (chips) with the other. Thus, the N_i are constant, but the Q_i may vary under the restriction

$$Q_1(t) + Q_2(t) = Q_1(0) + Q_1(0) = \text{constant.} \tag{8.1}$$

We then let the dual system $D_1 + D_2$ evolve until the new time-averaged distribution of each subsystem is established. Write the average energy of system 1 when it is in contact with system 2 as $\langle Q_{12}(t) \rangle$. According to our version of the equipartition principle, the energy is divided between the two systems in proportion to their numbers of atoms, so the mean energy of system

1 when in thermal contact with system 2 is

$$\langle Q_{12}(t) \rangle = \frac{N_1(Q_1 + Q_2)}{N_1 + N_2} = N_1 \langle q \rangle, \tag{8.2}$$

and, similarly,

$$\langle Q_{21}(t) \rangle = \frac{N_2(Q_1 + Q_2)}{N_1 + N_2} = N_2 \langle q \rangle. \tag{8.3}$$

For the purpose of generality we use subscripts i and j and define the *change* in mean energy content of system i as a result of thermal contact with system j as

$$\delta Q_{ij} \equiv Q_i(0) - \langle Q_{ij}(t) \rangle. \tag{8.4}$$

A little algebra gives

$$\delta Q_{ij} = \frac{N_j Q_i - N_i Q_j}{N_i + N_j}. \tag{8.5}$$

Exercise 8.8
Prove equation (8.5) and show $\delta Q_{ij} = - \delta Q_{ji}$.

From the definition of δQ_{ij}, it follows that if $\delta Q_{12} > 0$ then system 1 lost energy when it came into thermal contact with system 2. More specifically, the *mean* energy of system 1 was higher before thermal contact than after thermal contact. We say that system 1 was hotter than system 2.

Such changes in mean energy content provide a way of rank ordering the D_i. We shall call this a "temperature order" and shall adopt the following convention for the temperatures of the systems before the thermal contact was established:

If $\delta Q_{12} > 0$ then D_1 was hotter than D_2.

If $\delta Q_{12} = 0$ then D_1 and D_2 were at the same temperature.

If $\delta Q_{12} < 0$ then D_2 was hotter than D_1.

These permit us to establish a pair-wise ordering of the systems. This pair-wise scheme may be extended to a *complete* ordering by using the results of Exercise 8.9.

Exercise 8.9
Show that

$$\text{if } \delta Q_{ij} \geqslant 0 \text{ then } Q_i/N_i \geqslant Q_j/N_j. \tag{8.6}$$

Show that this leads to

$$\text{if } \delta Q_{ij} = 0 \text{ and } \delta Q_{jk} = 0 \text{ then } \delta Q_{ik} = 0; \tag{8.7}$$

$$\text{if } \delta Q_{ij} > 0 \text{ and } \delta Q_{jk} > 0 \text{ then } \delta Q_{ik} > 0. \tag{8.8}$$

Exercise 8.9 shows that if the temperatures of two systems equal the temperature of a third, they are equal to each other. Furthermore, if $T(D_1) > T(D_2)$ and $T(D_2) > T(D_3)$, then $T(D_1) > T(D_3)$. These results are sufficient to establish a complete ordering of the systems. It is a nonexclusive ordering, because two or more systems may occupy the same rank if they have equal temperatures.

(b) Setting Up the Quantitative Temperature Scale

The result of the previous section is known as the "zeroth law of thermodynamics." It permits us to define a thermometer and it guides us in setting up a quantitative temperature scale. We shall follow a line of argument that can be applied to a system of any size and that gives us the conventional definition of temperature when we consider a large system.

The state of an isolated system, D_i, is specified by N_i and Q_i. We look for a function $T_i(N_i, Q_i)$ that we can assign to each isolated system and that will serve the same ordering purpose as δQ_{ij}. That is, for each D_i we wish to assign a numerical value to $T_i(N_i, Q_i)$, and by examining the list of values we can put the systems in order. For example, if $T_i < T_j$, we say that system i has a higher temperature than system j, and we require that this same ordering would be achieved if we were to put the two systems into thermal contact and determine δQ_{ij} using the method we have already described.

If we could find such a function we would automatically obtain the desired relations:

$$T_i > T_j \text{ and } T_j > T_k \text{ implies } T_i > T_k,$$
$$T_i = T_j \text{ and } T_j > T_k \text{ implies } T_i = T_k.$$

Such a function $T_i(N_i, Q_i)$ provides a quantitative temperature scale. We set it up by considering two systems, D_i and D_j, that are in thermal contact and form the composite system D_s.

At this stage, we might proceed in either of two directions. We might discuss the operation and calibration of a gas thermometer in sufficient detail to provide a complete empirical definition of temperature. This would provide a link between the microscopic and macroscopic definitions of temperature, but it would require us to get ahead of ourselves. It must be based on a discussion of the kinetic theory of gases, a topic reserved for a later chapter.

To avoid this leaping ahead, we will proceed to a theoretical definition of temperature, based on the assumption that we know the energy content of a small system.

(c) Number of Configurations for a Composite System

We need to focus on two systems in thermal contact, having equal values of Q/N. We assume that the systems are large and recall the definition of $\Omega_i(N_i, Q_i)$ as the number of configurations available to a system of specified number of quanta and atoms. In Chapter 7, we derived (7.15), the approximation for large systems,

$$\ln \Omega_i \approx (N_i + Q_i) \ln (N_i + Q_i) - N_i \ln N_i - Q_i \ln Q_i \qquad (i = 1, 2, s)$$

We assume that systems 1 and 2 form a composite system s and we ask how the number of configurations available to the composite system is related to the numbers for the subsystems. In other words, what is Ω_s?

With some rearranging, we find for each subsystem

$$\ln \Omega_i \approx N_i \ln \left(1 + \frac{Q_i}{N_i} \right) + Q_i \ln \left(\frac{N_i}{Q_i} + 1 \right),$$

and we sum over the systems. Using the fact that Q_i/N_i is the same for all systems in thermal contact, we may take the terms in parentheses outside the summation:

$$\sum \ln \Omega_i \approx \ln \left(1 + \frac{Q_i}{N_i} \right) \sum N_i + \ln \left(\frac{N_i}{Q_i} + 1 \right) \sum Q_i,$$

$$\ln (\Omega_1 \Omega_2) = \ln \left(1 + \frac{Q_s}{N_s} \right) N_s + \ln \left(\frac{N_s}{Q_s} + 1 \right) Q_s$$

$$= \ln \Omega_s.$$

Thus the number of configurations available to the composite is the product of the numbers available to the subsystems,

$$\Omega_1 \Omega_2 = \Omega_s. \tag{8.9}$$

(d) Definition of the Theoretical Temperature

We now develop a definition that can be applied to systems of any size, although the form we choose is suggested by the properties of large systems. We start by taking the natural logarithms of both sides of the equation $\Omega_i \Omega_j = \Omega_s$, from which we obtain $\ln \Omega_i + \ln \Omega_j = \ln \Omega_s$. Now consider what would happen during thermal contact if a number of quanta δQ_i of energy ε were taken from system i and moved to system j. The energy changes in each system would be equal and opposite:

$$\varepsilon \, \delta Q_i = - \varepsilon \, \delta Q_j. \tag{8.10}$$

What would happen to the number of available configurations, Q_i, Q_j, and Ω_s? We have the relation $\delta \ln \Omega_i + \delta \ln \Omega_j = \delta \ln \Omega_s$, and if the systems are large the composite always has a value of Ω_s that is near the maximum. Thus the change of the total, $\delta \ln \Omega_s$, will be small and we can set it to zero, finding the approximate equilibrium condition,

$$\delta \ln \Omega_i = -\delta \ln \Omega_j. \tag{8.11}$$

Dividing equation (8.10) by (8.11) gives

$$\frac{\varepsilon \, \delta Q_i}{\delta \ln \Omega_i} = \frac{\varepsilon \, \delta Q_j}{\delta \ln \Omega_j}. \tag{8.12}$$

This relationship between energy fluctuations that occur during thermal contact will provide us with a definition of temperature. In fact, the function

$$T_i = \frac{\varepsilon \, \delta Q_i}{k \delta \ln \Omega_i}, \tag{8.13}$$

where k is a constant of proportionality, has the properties we seek. It can be evaluated for any system no matter how small, as long as we know how to compute the number of configurations $\Omega_i(N_i, Q_i)$. (The physical significance of T will depend on the size of the system, but that is another matter.) This definition of temperature has the nice property that the condition for thermal equilibrium is that the temperatures be equal: $T_i = T_j$.

Another property is useful. Putting $\delta Q_i = 1$ into the definition of T_i, we find

$$\frac{\varepsilon}{kT_i} = \delta \ln \Omega_i, \tag{8.14}$$

and for the change of the logarithm we have

$$\delta \ln \Omega_i = \ln \left(\frac{(N_i + Q_i)!}{(N_i - 1)!(Q_i + 1)!} \right) - \ln \left(\frac{(N_i + Q_i - 1)!}{(N_i - 1)! \, Q_i!} \right);$$

so after a little algebra,

$$\frac{\varepsilon}{kT_i} = \ln \left(\frac{N_i + Q_i}{Q_i + 1} \right). \tag{8.15}$$

This implies that T_i is a monotonically increasing function of Q_i/N_i. Therefore T_i has the required property of δQ_{ij} and it can serve as a temperature scale.

We have assumed Q to be large enough so that we can ignore the 1 in the denominator, and we can write

$$\frac{\varepsilon}{kT_i} = \ln\left(\frac{N_i}{Q_i} + 1\right)$$

or

$$\exp\left(\frac{\varepsilon}{kT_i}\right) = \frac{N_i}{Q_i} + 1. \tag{8.16}$$

This gives the following relation between the number of quanta per particle and the temperature:

$$\frac{Q_i}{N_i} = \frac{1}{\exp(\varepsilon/kT_i) - 1}. \tag{8.17}$$

This function is nearly a straight line when plotted. In fact, when the temperature is high and $kT/\varepsilon \gg 1$, we can expand the denominator and find $\varepsilon Q_i/N_i = kT_i$. Thus the coefficient k represents the high-temperature limit of the change in mean energy per particle for a unit change of temperature. It is also known as the Boltzmann constant. Its numerical value is connected to the numerical scales with which temperature and energy are measured.

8.4   ENTROPY

The entropy of a Boltzmann system D_i may be defined as k times the natural logarithm of the number of configurations that are allowed by the specified energy and number of particles in the system:

$$S_i \equiv k \ln \Omega_i(N_i, Q_i)$$

$$= k \ln\left(\frac{(N_i - 1 + Q_i)!}{(N_i - 1)! Q_i!}\right). \tag{8.18}$$

According to this definition, the entropy is a function only of N_i and Q_i, which are constant for an isolated system. As the reader can easily verify, the entropy increases with N_i and with the number of quanta Q_i.

In an isolated system, the values of N and Q are constant, and the time-averaged distribution of energy among the particles will be a weighted distribution. The weighting will be proportional to the likelihood of each distribution.

An important concept in thermodynamics is the change of total entropy that occurs when two systems are put into thermal contact, while the energy and number of particles remain constant. The results depend on the sizes of the systems and their initial temperatures. They may be summarized as follows.

Small Systems

(1) If two systems i and j, with entropies S_i and S_j, respectively, are put into thermal contact to form system s, the total entropy increases:

$$S_s > S_i + S_j. \tag{8.19}$$

(2) The amount of increase in total entropy is least when the two systems initially have equal energy per particle, $\varepsilon Q/N$.

Large Systems

(3) In the limit of large systems the rule becomes

$$S_s \geqslant S_i + S_j. \tag{8.20}$$

The equality holds if systems i and j are initially at the same temperature; otherwise the total entropy increases.

The proof of these rules is straightforward but involves messy calculations with factorials, so we will skip them and leave the reader to verify the rules through the following exercises.

Exercise 8.10

Consider two Boltzmann systems with equipartition: $N_1 = 30$, $Q_1 = 2$; $N_2 = 30$, $Q_2 = 2$. Show that $S_1 = S_2 = k \ln 465$, and when the systems are combined (i.e., brought into thermal contact), $S_{1+2} = k \ln 595665$, so the increase of entropy is $S_{1+2} - (S_1 + S_2) = 1.01k$.

Exercise 8.11

Suppose that one of the systems initially had all the energy, so the initial partitioning of energy was $N_1 = 30$, $Q_1 = 0$; $N_2 = 30$, $Q_2 = 4$. Show that the rise of entropy when the systems are brought into thermal contact is increased to $2.50k$. This verifies that a larger increase occurs when the systems are initially at different temperatures.

Exercise 8.12

Stirling's approximation is $\ln N! \approx N \ln N$, which is valid for large N.

(a) Use this two show that for large Boltzmann systems

$$S \equiv k \ln \Omega \approx kN \ln \left(1 + \frac{Q}{N} \right) + kQ \ln \left(1 + \frac{N}{Q} \right) \qquad (Q \gg 1, N \gg 1).$$

(b) Show that this leads, in the low-temperature limit, to

$$S \approx kQ + kQ \ln \left(\frac{N}{Q} \right) \qquad (Q/N \ll 1).$$

Thus, if two low-temperature systems with entropies S_1 and S_2 are put into thermal contact, the total entropy becomes.

$$S_{1+2} = k(Q_1 + Q_2) + k(Q_1 + Q_2)\ln\left(\frac{N_1 + N_2}{Q_1 + Q_2}\right).$$

Show that if $N_2/N_1 = Q_2/Q_1$,

$$S_{1+2} = k(Q_1 + Q_2) + kQ_1\ln\left(\frac{N_1}{Q_1}\right) + kQ_2\ln\left(\frac{N_2}{Q_2}\right) = S_1 + S_2,$$

so the increase of entropy is zero.

(c) Show that the increase of entropy also vanishes in the high-temperature limit $(Q/N \gg 1)$ if $N_2/N_1 = Q_2/Q_1$.

REFERENCES

Ciccotti et al. 1987
Crowford 1988
Ehrlich 1981
Eigen and Winkler 1983
Gould and Tobochnik 1988
Gurney 1949
Hill 1963
Kittel and Kroemer 1980
Mohling 1982
Reif 1967

PART TWO

PROCESSES IN CONTINUOUS SPACES

9

AN INTRODUCTION TO CONTINUOUS DISTRIBUTIONS

9.1 ON GUARD! A WARNING FROM BERTRAND'S PARADOX

In discussing events that occur in a *discrete* sample space, such as coin flips in which {H} and {T} are the discrete outcomes, we have little difficulty counting the number of possible outcomes. We open this chapter by illustrating a calculation involving a *continuous* sample space, in which an ambiguity can arise. It is possible to find apparently conflicting answers to probabilistic questions if the questions are not posed with sufficient precision. The apparent conflict in this case leads to the famous paradox first described by the French mathematician, Joseph Bertrand (1822–1900), in his book *Calcul des Probabilities.* Here is the problem in a slightly altered form, and it is illustrated in Figure 9.1,

> *Two concentric circles are drawn, one with half the radius (and hence one-quarter the area) of the other. A straight line, AB, is drawn at random through the outer circle, forming a chord. What is the probability that this chord will also intersect the inner circle?*

Three equally logical answers are possible, and they are 1/4, 1/3, and 1/2. Each depends on a different method of drawing the chord "at random." Because the method was not stated in posing the question, they are all acceptable, even though they give different answers:

(1) The center of the chord is considered to be at a random point in the outer circle, and this leads to the fraction 1/4, based on the relative areas of the two circles.

143

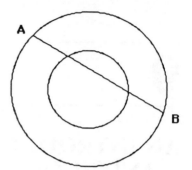

FIGURE 9.1   Construction for Bertrand's paradox. The outer circle has twice the radius of the inner. What is the probability that a random chord *AB* of the outer circle will intersect the inner?

(2) One end of the chord is fixed while the other end is taken as a random point on the outer circle. This leads to the fraction 1/3, because the inner circle subtends an angle of 60° from the fixed end of the chord.

(3) The random lines are all drawn parallel to each other. Of these parallel lines, one-half will intersect the inner circle because it has one-half the radius of the outer circle.

In fact, all three answers are acceptable because the problem was poorly specified. The ambiguity arises because we deal with a continuous sample space and there are an infinite number of possible outcomes. The computed fraction depends on the method of establishing the set of random chords, and in addition to the three solutions proposed by Bertrand, there are countless other solutions. (For example, a larger fraction of hits occurs if the chords are defined by a pair of randomly selected points.)

Exercise 9.1
Develop a computer program to evaluate the answers to this problem by simulating the random selection of lines by each of the three methods. Tabulate a large number of results and determine how many trials are necessary to show convincingly that they give distinct answers.

A helpful insight into this problem can be obtained by a formulation that is closer to the original formulation by Bertrand. Imagine a triangle with three equal sides, *L*, to be inscribed inside the larger circle. (It is left as an exercise for the reader to show that $L = 2\sqrt{3}$.) Bertrand posed the problem as follows:

> *If a chord of the outer is drawn at random, what is the chance that it will be shorter than the sides of the inscribed equilateral triangle?*

It is not difficult to show that the sides of the inscribed triangle will be tangent to the inner circle. This means that all chords longer than *L* will intersect the inner circle while all chords shorter than *L* will not. So the problem may be solved by finding the distribution of chord lengths, *N(L)*, generated by each

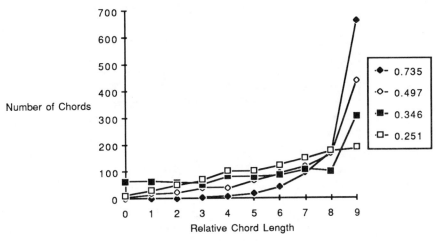

FIGURE 9.2   The distribution of chord lengths generated in a unit circle by four different geometric constructions. The puzzle posed by Bertrand was to find the fraction of chords with lengths greater than a critical value. The solution depends on the method of construction and this leads to the "paradox" of multiple solutions. See text.

method of construction and counting the number of chords that are longer than L. This approach is illustrated in Figure 9.2, which shows the distribution of chord lengths for four methods of construction.

The chords were divided into ten length categories, 0–9, ranging from length zero to the maximum, which is the diameter of the outer circle. In all four constructions, the longer chords were more frequent, but the shape of the distribution depends strongly on the method.

For each method, the number of chords that intersected an inner circle of radius one-half was counted, and the fractions are indicated by different symbols. In order of decreasing fraction, the methods (and their theoretical fractions) are:

(1) Two random points define the chord (0.7788).
(2) Randomly chosen points on a single diameter define the centers of chords (1/2).
(3) Two randomly chosen points on the circumference define endpoints of chords (1/3).
(4) One random point defines the center of each chord (1/4).

As a fifth example, it is possible to show by elementary methods that if chords were constructed with a uniform distribution of lengths, the fraction $(4 - 2\sqrt{3})/4 = 0.134$ would intersect the inner circle.

Another famous probability problem concerning continuous sample spaces is illustrated in Figure 9.3, the Buffon needle problem. Its analysis is left to the

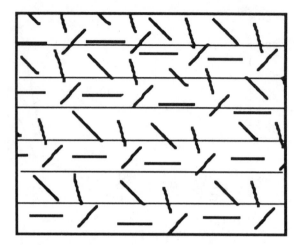

FIGURE 9.3   The Buffon needle problem. If the separation of the lines equals the length of the needles, what fraction of randomly tossed needles will touch a line? (The needles are assumed to be randomly oriented and to have their central points at random locations.) This is an example of a well-posed problem that can be solved by Monte Carlo simulation. A bit of calculus shows that the reuslt can be used to evaluate π.

reader, who might also perform the real experiment by dropping a needle onto a ruled surface.

Exercise 9.2

In *Scientific American* for October 1959, pp. 174ff., Martin Gardner describes Bertrand's paradox (discussed above) as well as the following problem of the broken stick. If a stick is broken randomly into three pieces, what is the probability that the pieces can be put together into a triangle. (The condition is that the sum of the lengths of the two shorter pieces must exceed that of the longest piece.) The answer depends on the method of breaking. Carry out a simulation to verify that $p = 1/4$ if the stick is broken at two points randomly chosen on the original stick; but $p = 1/6$ if the stick is first broken into two at a randomly selected point and then one of the two pieces is randomly selected and broken.

9.2   CONTINUOUS DISTRIBUTIONS

(a) Cumulative and Density Distributions

Measured quantities often take a continuous set of values rather than a discrete set. The distance of a point from one end of a line is an example, and the speed of a free atom is another. The distribution of this type of quantity requires a slightly different approach, because we cannot count the number of points in a line. Instead, we use the concept of relative measure as follows.

Consider a point at a distance x from the end of a line whose length is A. The fraction of the line that lies between the origin and the point is x/A. This

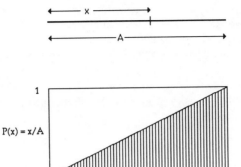

FIGURE 9.4 The cumulative probability of a random point in the line falling to the left of x is $P(x) = x/A$.

means that if a second point is chosen at random it will have the probability $P(x) = x/A$ of lying between the first point and the origin (Figure 9.4). Capitalized $P(x)$ is the cumulative probability function. It is a continuous function of distance, x, and it varies from zero to unity as we move from one end of the line to the other: $P(0) = 0, P(A) = 1$.

Having located the first point at x, let us now ask for the probability that a second point located at random will fall exactly on the first point, at x. Clearly the answer is zero, because the point has zero measure, and yet this cannot be the entire story because the second point must fall somewhere. We need a means of coping with the fact that points have zero measure relative to the line.

We consider a short interval, $(x, x + dx)$, along the line, and instead of trying to specify how many points there are in dx, we write the probability of falling in the interval $(x, x + dx)$ as $p(x) dx$ and set it equal to the ratio of the measures, dx/A, so

$$p(x)\, dx = \frac{dx}{A}. \tag{9.1}$$

This vanishes when dx vanishes. We can also write it as the difference of the cumulative probabilities,

$$P(x + dx) - P(x) = \frac{dx}{A}$$

$$= p(x)\, dx. \tag{9.2}$$

Rewriting the left-hand side terms of the derivative of P, we have

$$\frac{dP}{dx}\, dx = p(x)\, dx. \tag{9.3}$$

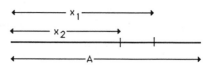

FIGURE 9.5 Two points are placed at random. What is the probability that the second will lie to the left to the first?

Thus the probability density, $p(x)$, can also be defined as the derivative of the cumulative probability,

$$\frac{dP}{dx} \equiv p(x). \tag{9.4}$$

With this definition, the probability that a point randomly chosen on a line will fall in the interval $(x, x + dx)$ is just $p(x)\,dx$, and for a uniform density, $p(x) = 1/A$. The probability density is the factor of proportionality between the probability of falling in the interval $(x, x + dx)$ and the width of the interval, dx. It does not vanish when dx shrinks to a point, although the probability of falling directly on a point, $p(x)\,dx$, does vanish.

Of course, the probability density will not always be a constant, but it must always satisfy the normalization condition, namely, that the sum over all intervals in the sample space must be unity. We may express this as the integral over the length of the line, $\int p(x)\,dx = 1$.

As an example of the use of the probability density function, let us suppose that two points (at x_1 and x_2) are placed at random on a line of length A and ask for the probability that the second will lie closer to the origin than the first. We break the solution into two steps (Figure 9.5).

First, we assume that x_1 falls in the interval $(x, x + dx)$ with probability dx/A, and we calculate the conditional probability that $x_2 < x_1$ given $x_1 = x$. This is the product $(x/A)(dx/A)$. Next, we sum over all possible positions for the first point by carrying out the integration

$$P(x_2 < x_1) = \int \left(\frac{x}{A}\right)\left(\frac{1}{A}\right) dx = \frac{1}{A^2} \int x\,dx = \frac{1}{2}. \tag{9.5}$$

Exercise 9.3
Suppose two points are randomly placed on a line of length A. What is the probability that a third randomly placed point will fall between them?

(b) Converting from Continuous to Discrete Distributions

Mathematically speaking, the possible speeds of atoms in a gas make up a continuum. If two atoms have been measured, it would be possible, in principle, to find an atom whose speed is intermediate between them. But the number of visible atoms is finite and there is a practical limit to the precision with which we can measure the speed of an atom. Thus it is often convenient to divide the

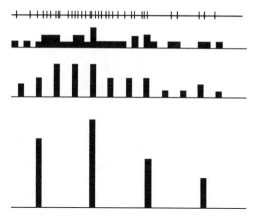

FIGURE 9.6   Schematic distribution of speeds along a continuous line (upper). In the lower three histograms, the speeds have been assigned to bins of different sizes to illustrate the influence of bin size on the apparent structure of the data. The middle selection is the most useful in revealing the data.

speed scale into a discrete set of intervals and count the numbers of atoms in each interval. We pretend that each atom in a bin has exactly the speed that is assigned to the bin.

The selection of bin size is a matter of compromise. If the bin is too broad, the features of the distribution will be blurred; if it is too narrow, many of the intervals will be empty and the placement of individual atoms will be affected by errors of measurement. Ordinarily, the interval is chosen to be at least as wide as a typical error of measurement. In effect, we are creating a discrete distribution by placing the atoms into a finite number of speed bins. We call this "binning" the data.

Figure 9.6 is an example of selecting three different bin sizes. The upper portion shows the speeds of 34 atoms represented on a continuous speed line, and the lower histograms show them assigned to bins of various sizes. The effect of improper choice of bin size is obvious. (As Goldilocks said when the examined the chairs of the three bears, one is "too large," one is "too small," and one is "just right" to reveal relevant structure in the data.)

Not only must the *interval size* be chosen to match the precision of the speed measurements and the number of atoms in the sample, but the *nature of the interval* must be clearly specified. Depending on the choice, the appearance of the distribution may vary considerably, and if the nature of the interval is not appreciated, the distribution may be misinterpreted.

For example, suppose we consider a small cell of the earth's atmosphere and imagine that all the atoms in that cell have the same speed but are moving in random directions. We wish to represent the number of atoms moving in different directions relative to the zenith (straight upward). Let us compute what to expect in 10-degree zones of zenith distance, $(D, D + 10)$. We wish the number

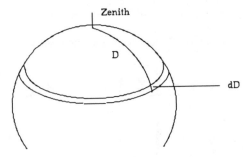

FIGURE 9.7 Computation of the areas of zones on a sphere of unit radius at a constant distance from the zenith. D is the distance from the zenith and dD is the width of an infinitesimal zone. The circumference of each zone is $2\pi \sin D$, and its area is $(2\pi \sin D)\,dD$.

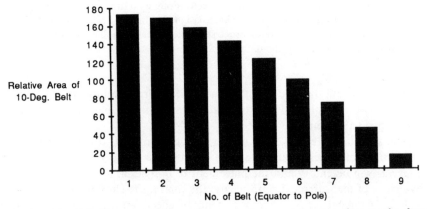

FIGURE 9.8 This histogram shows the nonuniform appearance that results from plotting a uniform distribution of the directions in terms of distance from the zenith (pole). It illustrates the importance of carefully choosing the spacing in establishing a discrete distribution from a continuous one. The distribution appears nonuniform because the bins of higher number (near the zenith) correspond to smaller areas. The area of a belt whose sides are a distances D_1 and D_2 from the zenith is proportional to $2(\cos D_1 - \cos D_2)$. Hence a division according to equal intervals of the cosine of the distance from the zenith would give equal areas and would give a more uniform appearance to the histogram.

of atoms, $N(D, D+10)$ in each 10-degree belt. This is proportional to the belt areas, $2\pi \sin(D)\,dD$ (see Figure 9.7). So we have

$$N(D, D+10) = 2\pi \int_{D}^{D+10} \sin(D')\,dD', \qquad (9.6)$$

and this gives

$$N(D, D+10 = 2\pi[\cos(D) - \cos(D+10)]. \qquad (9.7)$$

Figure 9.8 shows a schematic histogram resulting from this equation, constructed for an idealized gas in which the speeds of the atoms are distributed uniformly in all directions. The histogram shows the numbers of atoms in contiguous 10-degree zones around the zenith and the number decreases toward the zenith due to the decreasing areas of the zones.

REFERENCES

Gardner 1959
Mosteller et al. 1970
Rubinstein 1981

10

EXPONENTIAL PROCESSES

This chapter formulates the simulation of two simple decay processes: radioactive decay and photon absorption. These processes occur on a continuous space (a time-line or a one-dimensional space), but when we simulate them with a digital computer, we must represent them in a discrete space.

10.1 RADIOACTIVE DECAY

Suppose a box contains N radioactive particles at the initial time, $t = 0$. We continuously monitor the box with a geiger counter and note the times, t_i, at which particle decays occur. Examining these data, we cannot predict when the next decay will occur, but a pattern in the frequency of decays begins to emerge. More particles decay early and few decay later; in fact, it is possible to define an analytical function to describe the average behavior of the particles.

We assume that all particles consist of the same nuclear species, so they are all described by the same statistics, and we describe the decay process in two distinct, but equivalent, ways.

(a) First Representation: Decay Events

We suppose that a particle has a probability $p = r\, dt$, of decay in the short time interval dt. This implies that the probability is proportional to the time interval, and our task is to derive the behavior over longer, finite intervals. The number r is positive and it has the units of inverse time (1/second), because the probability p is a pure number and it lies between 0 and 1. In the limit as $dt \to 0$, the

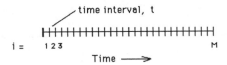

FIGURE 10.1   Division of the time-line into small intervals δt for simulating radioactive decay. Each interval is labeled with the index $i(1 < i < M)$. An unstable particle has a probability $p = r \, \delta t \ll 1$ of decaying in each time interval. The goal is to determine the distribution of decay events along the time-line.

probability of decay will vanish. For simplicity, we assume r to be the same for all particles in the box.

In order to simulate this process with a digital computer, we imagine the time-line to be finely divided into equal intervals of width δt, sequentially labeled with an index, $i = 1, 2, \ldots, M$, where M is the total number of time steps. (See Figure 10.1.) The beginning of each interval is $t_i = i \, \delta t$. When δt has been selected, the constant probability $p = r \, \delta t \ll 1$ may be evaluated and stored. It must satisfy the inequality $p \ll 1$, meaning that the interval is so short that the probability of decay is small.

We consider the history of each particle, $j = 1, \ldots, N$, individually, indexing through j. For each particle, the program initializes the clock to $i = 0$ and then indexes forward through time $(t_i \to t_{i+1})$, and in each interval a random number is drawn to decide whether the particle has decayed. This random number, x, is drawn from a distribution that is uniform in the interval $0 < x < 1$. If $x < p$, the particle is said to decay in the ith time interval. When the particle decays,

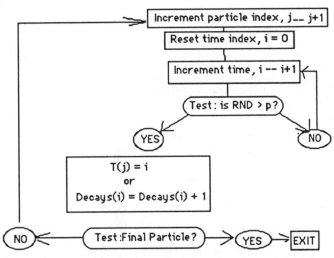

FIGURE 10.2   Flowchart of program section for keeping track of decay events, either by noting the individual times of the events, $T(j)$, or by incrementing the counters in cells, DECAYS.

we enter the time-index i into an array $T(j)$ at the location of the jth particle. Then set the clock back to $t = 0$, and go on to the next particle, $j \to j + 1$, and step the clock forward to follow the next particle's history. A schematic flow-chart to describe this program is shown in Figure 10.2.

This method of describing the process has a weakness that becomes apparent if we merely display the decays as marks along a time-line at each $T(j)$. It is difficult to see the pattern, and there may be two decays in the same time interval. A histogram of the distribution of endpoints would be more informative. This can be achieved by establishing an array, DECAYS, containing the number of particles that decayed in each time interval. To achieve this, we merely replace the statement $T(j) = i$ with the statement

$$DECAYS(i) = DECAYS(i) + 1.$$

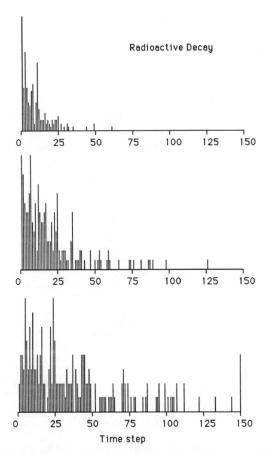

FIGURE 10.3   Simulated radioactive decay events plotted as a function of time.

A plot of the array DECAYS(i) against i then gives a histogram of the frequency of decay as a function of time. The results for three representative calculations are shown in Figure 10.3. As expected, the processes with smaller p (lower portion of figure) are spread more widely on the time-line. In fact, the width of the spread is proportional to $1/p$, so this number (which has the units of time) provides an estimate of the average lifetime of the particle. We shall prove this result later in this chapter.

A striking feature of these diagrams is that the mean frequency of the decay events decreases with time. At any time, the expected frequency is proportional to the number remaining, and this decreases with time.

(b) Second Representation: Flux

In place of focusing on the instants of the decays, we now consider the number of particles that persist to each time-step. The resulting calculation contains the same information as the preceding, but in a slightly different form, since this curve is the integral of the first. (The calculations also differ slightly in their efficiency.)

Define the array FLUX(i) as the number of particles that have not decayed by interval i. Initialize all its cells to 0. Then take the first particle and step through the time intervals $i = 1, 2, \ldots$, testing for decay as before. If the particle does not decay in that interval, increment the corresponding cell of FLUX(i) by unity. If the particle does decay, stop with that particle, reset the time index to $i = 0$, and restart with the next particle. Repeat this process until all particles have been treated. Another way to describe the process is to increment the successive cells of FLUX(i) out to the i corresponding to the particle decay. Do this for each particle. The accompanying flowchart (Figure 10.4) describes this procedure.

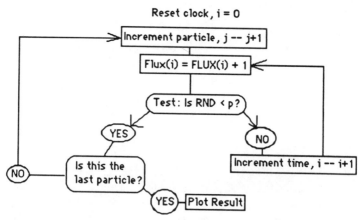

FIGURE 10.4 Flowchart for simulating the number of particles that remain undecayed at each time-step.

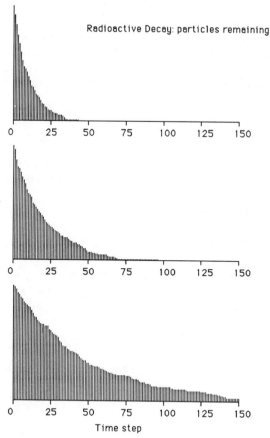

FIGURE 10.5 Particles remaining during a radioactive decay simulation. Compare with Figure 10.3.

The result is an array in which FLUX(0) = N, and successive entries are smaller. A representative calculation is shown in Figure 10.5 based on the same parameters as Figure 10.3.

The two curves are similar in average shape but they differ in roughness. The FLUX histogram is built by a series of incrementing cells for each particle until the DECAY of that particle occurs. Therefore each DECAY event represents a step-decrease in the FLUX. Or to put it more formally, the DECAY curve is the derivative of the flux curve.

Readers familiar with calculus may recognize this proportionality between a function and its derivative as an earmark of the exponential function. (The exponential $f = e^{ax}$ has the derivative $df/dx = ae^{ax} = af$.) This similarity suggests that the exponential function describes this decay process, and we could, in fact, directly solve this problem with calculus: By writing down a differential equation for the number of decays and integrating it analytically, we could find an exponential function as the solution. Rather than do that, and in order to clarify the nature of Monte Carlo simulations in probability, we will proceed with a more elementary analysis that introduces the definition of the exponential and involves only algebra. Before doing so, we shall introduce a physically distinct process that is mathematically identical to the radioactive decay process.

10.2   PHOTON ABSORPTION

In order to discuss a similar problem, relating to the absorption of light in gas, we enter for the moment into the two-dimensional "flatland" of Figure 10.6. This diagram represents the cross section of a slab region containing absorbing particles of cross-sectional area σ. Imagine that N photons arrive in a parallel stream from the left. We wish to calculate how many photons will emerge from the other side of the slab. The slab is dx units thick and dy units high and there are n particles per unit area. The total number of particles is $n\,dx\,dy$, and each particle creates a shadow, so the total shadowed area is $n\sigma\,dx\,dy$, as long as $n\sigma\,dx$ is not too large and the particles do not cast shadows on each other. Therefore the fractional shadowed area is $n\sigma\,dx$, and we may write

$$\text{Probability of being absorbed} = n\sigma\,dx. \qquad (10.1)$$

This probability is similar in form to the probability of radioactive decay described earlier. If we consider a sequence of such slabs, the problem is

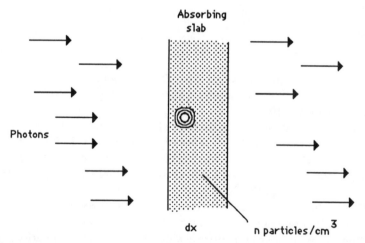

FIGURE 10.6   Photon absorption in a thin slab containing n particles/cm^3, each with a cross section of σ cm^2.

Two identical absorbing slabs

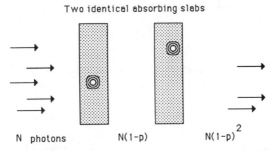

N photons   N(1-p)   N(1-p)^2

FIGURE 10.7 Absorption to a sequence of identical slabs. The probability of absorption in each slab is $1 - p$, so the probability of sucessive transmission through M slabs is $(1 - p)^M$. Tracing the transmission of photons through such a series of slabs is similar to tracing the histories of radioactive particles.

mathematically identical to that of radioactive decay, and the slabs take on the role played by the time intervals. Figure 10.7 illustrates the geometry for this problem, and each slab is assumed identical to the others; that is, the particle density n is a constant. The simulation of this process would be identical to that of the radioactive decay, with the spatial dimension perpendicular to the slab (or along the flight of the photons) taking the place of time. An identical computer program could be used with a redefinition of the variables. We now turn to the algebraic analysis, as an illustration of the mathematical analysis of such processes.

Exercise 10.1
Develop a computer program to carry out this simulation in both versions and compare their speeds.

10.3   ANALYSIS OF PHOTON ABSORPTION

We may derive an analytical expression for the solution of the photon absorption problem—and one that also applies to the radioactive decay—as follows. The probability of transmission through the first slab is

$$T(1) = 1 - p. \tag{10.2}$$

If the transmissions through the two slabs are statistically independent events (see Appendix 1), the probability of being transmitted through two slabs may be written as the product of the probabilities for each slab considered singly; that is, $T(2) = T(1)T(1) = (1 - p)^2$.

We can at once generalize our result and write for the probability of transmission through M such layers:

$$T(M) = T(1)^M = (1 - p)^M. \tag{10.3}$$

By combining this with the transmission of a second series of N identical layers, we have

$$T(M + N) = T(1)^M T(1)^N = T(M)T(N). \tag{10.4}$$

Thus, if the transmission of a slab of thickness M is $T(M)$ and of thickness N is $T(N)$, the transmission of a slab of thickness $M + N$ is the product $T(M)T(N)$, as long as the events are statistically independent. The function $T(y)$ therefore has the characteristic behavior of the *exponential function*, which is defined so that the product of the functions at two values of the argument, M and N, is equal to the function evaluated for the sum of arguments, $M + N$.

We can say a bit more about the function $T(y)$ without a detailed analysis. We known $T(0) = 1$, because all photons will penetrate a "slab" of zero thickness. Also, in the limit of thick slabs, the transmission must vanish. Furthermore, we see on physical grounds that $T(y)$ must be a monotonically decreasing function of y, because any increase of thickness will lead to a decrease in the transmission probability.

If we set out to compute a table representing the function $T(y)$, we might start by imagining a series of n identical slabs, whose transmission is

$$T(1)^n = T(n). \tag{10.5}$$

This means that if we know the value of $T(1)$, we can calculate $T(n)$ by raising $T(1)$ to the power n.

So our task is to compute $T(1)$, and when we have done this, the function will be defined. We start by reverting to the original description of the (small) probability of absorption in a single thin slab, $p = n\sigma\, dx$. Because this must be infinitesimal if dx is infinitesimal, we write it with the infinitesimal symbol, $d\tau = n\sigma\, dx$. Then the probability of transmission through a thin slab is $T = 1 - d\tau$.

Now suppose there is a uniform, finite slab of thickness X. With this slab we associate the finite quantity $\tau = n\sigma X$, which we call the "optical thickness" of the slab. We do not know the transmission of this finite slab, but we can compute it with a simple trick. We imagine it divided into a very large number, N, of equal slabs of thickness $dx = X/N$ so $d\tau = \tau/N$. Then we may write the probability of transmission through all the slabs as the product of a very large number of terms; that is,

$$T(\tau) = \lim_{N \to \infty} \left(1 - \frac{\tau}{N}\right)^N. \tag{10.6}$$

We are interested in the special case where $\tau = 1$. This is given by the expression

$$T(1) = \lim_{N \to \infty} \left(1 - \frac{1}{N}\right)^N. \tag{10.7}$$

The right-hand side may be evaluated with a pocket calculator searching for the limiting value by substituting larger and larger values of N. For example, with $N = 32$, we find $T(1) = 0.36201$, and as N becomes immense we gradually approach the value $T(1) = 0.36788$. This solves our problem, because given $T(1)$, we can find $T(\tau)$ from the expression

$$T(\tau) = T(1)^N. \tag{10.8}$$

We can save some work by noting that the general expression for $T(\tau)$ is the definition of the exponential, which may therefore be written

$$T(\tau) = \lim_{N \to \infty} \left(1 - \frac{\tau}{N}\right)^{-N} = e^{-\tau}. \tag{10.9}$$

Thus the transmission can be expressed as $T(\tau) = e^{-\tau}$, where $\tau = n\sigma X$ is known as the "optical thickness." Using tabulated or computer-generated values of the exponential function, we can compute the transmission probability for a finite slab of arbitrary thickness.

Exercise 10.2
Demonstrate the exponential behavior of transmission function $T(M + N) = T(M)T(N)$ as follows. First simulate the transmission of photons through two slabs whose optical thicknesses are 0.25 and 0.75, respectively, and evaluate the two transmission probabilities. Then simulate the transmission through a single slab of optical thickness 1.0. Show that this transmission probability equals the product of the other two, whose total optical thickness is also 1.0.

10.4   MEAN TRANSMISSION PATH

Suppose we know p, the probability that a photon will be absorbed in a single slab. If p is not too small, we know that any particular photon has a good chance of penetrating several such slabs before being absorbed. If we set up a series of slabs and number them, we may ask for the average slab number in which the absorption will take place. We call this the "expectation value of the slab number" and indicate it with brackets, $\langle N \rangle$.

We start from $P(N)$, the probability of being absorbed in the Nth slab. This is given by the probability of passing through $N - 1$ slabs and being absorbed in the next, or

$$P(N) = (1 - p)^{N-1} p. \qquad (10.10)$$

The expectation value of N may be found by a weighted average of all the possible outcomes, each outcome weighted by its probability, so

$$\langle N \rangle = \sum_{N=1}^{\infty} N P(N), \qquad (10.11)$$

or

$$\langle N \rangle = \sum N p (1 - p)^{N-1}$$

$$= p + 2p(1 - p) + 3p(1 - p)^2 + \cdots.$$

This may be written

$$\frac{\langle N \rangle}{p} = 1 + 2(1 - p) + 3(1 - p)^2 + \cdots$$

or

$$\frac{\langle N \rangle (1 - p)}{p} = 1 - p + 2(1 - p)^2 + \cdots.$$

Subtracting the second of these from the first, and using the summation of a geometric series, we have

$$\langle N \rangle = \frac{1}{p}. \qquad (10.12)$$

Thus the photon is expected to penetrate a number of slabs given by the reciprocal of the probability of absorption in a single slab. Looking back on the radioactive decay simulation, we see that this is similar to the result that emerged there, namely, that the particles survived for a time that was proportional to $1/p$.

Exercise 10.3
Verify the relationship of the mean path to the probability with your simulation program.

10.5   MIXED PHOTONS AND THE FAILURE OF STATISTICAL INDEPENDENCE

We have just seen that, on certain assumptions, the probability of transmission through a series of slabs is the product of the individual probabilities. This is the multiplication rule for the probabilities of statistically independent events.

The multiplication rule led us to the exponential law of attenuation, so when the multiplication rule is not valid, we should not expect the exponential rule to hold. We now illustrate a condition in which the multiplication rule fails.

Looking at the derivation of the multiplication rule in Appendix 1, we see that it requires $P(B|A) = P(B)$. the right-hand side of this equation is the probability of passing through B, and the left-hand side is the probability of passing through B after having passed through A. the equality of these two implies that the passage through A can have no influence on the probability of passage through B. In other words, nothing happens to the group of photons while passing through A that could affect the likelihood of their passage through B. This is an intuitively valid condition when we deal with a single photon, because the photon will either be transmitted or absorbed by the slab. If it is transmitted, it is assumed to be unchanged, and its probability of transmission through B is not affected.

So we expect the exponential law to apply to single photons, or to a beam of identical photons. But if we apply this to a heterogeneous beam of photons of different wavelengths, this will not be the case. In general, photons of different wavelengths will have different values of p, because the atoms in the slab will absorb some colors more effectively than others. Suppose we have a beam consisting of N_1 photons of one type and N_2 of a second type. The photons pass through slab A and we measure their numbers, finding N_1' and N_2'. (See Figure 10.8.)

Suppose the measured numbers imply different transmission probabilities in slab A for each type:

$$T_1(A) = \frac{N_1'}{N_1} \quad \text{and} \quad T_2(A) = \frac{N_2'}{N_2}. \tag{10.13}$$

FIGURE 10.8   Measuring the transmission probability of two types of photon passing through two slabs, A and B. The probabilities are evaluated from the fractions that pass through each slab. The photons of each type separately obey the exponential law, but the combined beam does not, because the passages through the two slabs are not statistically independent events when the photons are considered to form a single beam. The first slab alters the composition of the beam.

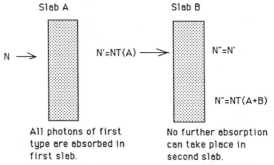

Slab A   Slab B

$N \rightarrow$ $N' = NT(A) \longrightarrow$ $N'' = N'$

$N'' = NT(A+B)$

All photons of first   No further absorption
type are absorbed in   can take place in
first slab.   second slab.

FIGURE 10.9   When a composite beam is passed through the two slabs, the behavior of each type of photon will be different, and the transmission probability will not obey the multiplication rule that applies to statistically independent events. In this case $T(A + B) = T(A)$ because the second slab can have no effect on the beam; all the photons of the first type were removed in the first slab, and the second slab lets this type pass through.

Similarly, their numbers are measured after passing through slab B, giving N_1'' and N_2'', and the corresponding transmission probabilities:

$$T_1(B) = \frac{N_1''}{N_1'} \quad \text{and} \quad T_1(B) = \frac{N_2''}{N_2'}. \tag{10.14}$$

These are all the data we need to demonstrate that statistical independence does not hold in the case of mixed photons.

Define transmission probabilities $T(A)$ and $T(B)$ for the entire beam, as though our measuring apparatus could not distinguish between the two colors of photons. They are given by

$$T(A) = \frac{N_1' + N_2'}{N_1 + N_2}, \tag{10.15}$$

$$T(B) = \frac{N_1'' + N_2''}{N_1' + N_2'}. \tag{10.16}$$

Also define $T(B|A)$ as the transmission probability for photons in slab B after they have passed through slab A. If we measure the transmission probability for the pair of slabs, it is $T(A + B) = T(B|A)T(A)$. But $T(B|A) \neq T(B)$, so $T(A + B) \neq T(A)T(B)$.

That is, *with mixed photons the transmission does not obey the multiplication rule*, because the passage of the beam through the second slab is influenced by its having passed through the first slab (Figure 10.9).

Exercise 10.4

Extend the radioactive decay simulation to include two species of particle with decay times that differ by a factor of 3. Simulate a large number of decays and see how the composite mean curve of radioactivity departs from an exponential curve. (Save this program and its results. Later in this book, we develop a technique for inverting this problem and determining the decay times from the composite curve of decay activity.)

REFERENCE

Reif 1967

11

THE POISSON DISTRIBUTION

There are two distributions (Poisson and normal) that are approached by the Bernoulli distribution (described in Chapter 4) in the limit of large numbers of trials. Under appropriate conditions, these limiting expressions simplify many numerical calculations, and they also greatly simplify the mathematical analysis. In this chapter, we examine the Poisson distribution, which is appropriate when p, the probability per trial, is very small while the number of trials is large. In Chapter 12, we examine the "normal" or Gaussian distribution, which holds for any value of p.

We shall derive the Poisson distribution in two ways. The first derives the Poisson distribution directly from the description of a stochastic process; the second displays it as a limiting case of the Bernoulli distribution and as a handy approximation avoiding the need to compute the factorials of large numbers.

11.1 DESCRIPTION OF THE POISSON PROCESS FOR A RADIOACTIVE CHAIN

As an example of a Poisson process, we will consider a chain of radioactive decays, in which a series of nuclear species S_n ($n = 0, 1, 2, 3, \ldots$) decay from one to the next. Figures 11.1a and 11.1b show the results of a simulation of the decay of such a hypothetical series of species. One hundred nuclei start in state 0. Each is assumed to decay to the next step with a probability $\mu = \frac{1}{50}$ per year. When the next step has been achieved, the nucleus has the same probability of decaying to the successive steps. The simulation was performed in 1-year

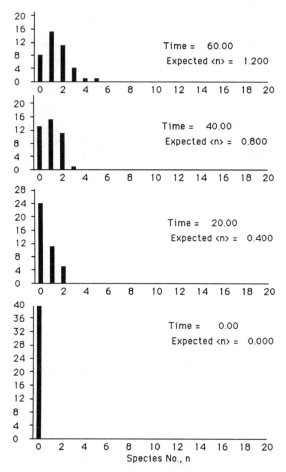

FIGURE 11.1a   Snapshots of species populations $N_n(t)$. Time increases up.

intervals, and the histograms show the number of each type of nuclei as time progressed.

To express the problem formally, we will trace the history of each nucleus by saying that it is in state n if it has taken n steps down the chain. Note that n cannot decrease, and therefore the following rules hold:

$n \geqslant 0$,   that is, n is 0 or a positive integer;

$n \rightarrow n + 1$,   that is,   it can only increase.

Each nucleus has an equal probability, $p(n \rightarrow n + 1)$, of moving ahead at each time-step, δt. We write this probability as

$$p(n \rightarrow n + 1) = \mu \, \delta t. \tag{11.1}$$

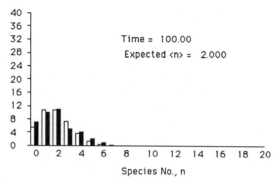

FIGURE 11.1b A single snapshot similar to Figure 11.1a, comparing the theoretical expectation (hollow) and the simulated (solid) results at time = 100 and $\langle n \rangle = \mu t = 2$.

It is proportional to the small time interval δt, with a constant of proportionality μ, which has the units *per unit time*. The probability $p(n \rightarrow n + 1)$ vanishes as $\delta t \rightarrow 0$. The sum over all possibilities is unity:

$$\sum_{j=0}^{\infty} p(n \rightarrow n + j) = 1. \tag{11.2}$$

The Poisson process is defined by the rule that n can only change by increments of 0 or $+1$:

$$p(n \rightarrow n + 1) = \mu \, \delta t,$$
$$p(n \rightarrow n + j) = 0 \quad \text{for all } j \neq 0, 1. \tag{11.3}$$

This implies that the probability that a nucleus will *not* decay is

$$p(n \rightarrow n) = 1 - p(n \rightarrow n + 1) = 1 - \mu \, \delta t. \tag{11.4}$$

Let $P_n(t)$ be the probability that a nucleus will be in state n at time t. The distribution $P_n(t)$ is called the Poisson distribution. We may ask how it evolves with time for a given n, as in Figure 11.1a, or we may ask how it varies with n for a given t, as in Exercise 11.1.

Exercise 11.1
Carry out a simulation in which $p(n \rightarrow n + 1) = \delta t / 10$ in each step. Use $\delta t = 0.1$ and continue for 1000 steps, until $t = 100$. Follow the histories of many particles, and derive the probability distribution among the final states, $P_n(100)$.

To find out what we ought to expect, let us start with $P_0(t)$, the probability that a particular nucleus remains in its original state after time t. At each step

δt the probability of not decaying is

$$p(0 \to 0) = 1 - \mu \, \delta t. \tag{11.5}$$

We let the finite time interval, t, be divided into a large number, N, of short intervals, δt, and we focus on the behavior when N becomes large. After $N = t/\delta t$ steps, the nucleus has the probability

$$P_0(t) = (1 - \mu \, \delta t)^N$$
$$= \left(1 - \frac{\mu t}{N}\right)^N$$

of remaining in its original state. For very large N (i.e., small δt) this may be rewritten using (10.9) as

$$P_0(t) = \exp(-\mu t), \tag{11.6}$$

which, as expected, decreases with the passage of time. Thus the probability of remaining in the original state decreases exponentially, and we note that

$$\mu t = \langle n \rangle \tag{11.7}$$

is the expected number of steps that occur in time t.

Note the similarity of this process to the absorption of a photon, which led to the exponential distribution in Chapter 10. In that case, we said that the traveling photon has the small probability $\kappa \, \delta x$ of being absorbed in each space interval δx and probability $1 - \kappa \, \delta x$ of surviving the interval. We only permitted the photon one state of existence; once it has been absorbed it no longer existed. In the present case, the nuclei are transformed from one state to the next. The original state starts with unit probability and decays exponentially with time because nuclei cannot return from higher states. The higher states start with zero probabilities and they obey a different distribution, as we shall now show.

11.2   DIFFERENTIAL EQUATION LEADING TO THE POISSON DISTRIBUTION

We now establish a set of differential equations for the probability of the nth state as a function of time, $P_n(t)$. The solution of these equations by elementary calculus then leads to the Poisson distribution. (By a similar method, we established the differential equation leading to the exponential distribution in Chapter 10).

The Poisson process is illustrated schematically in Figure 11.2. We may write equations for the change of $P(n, t)$ with time as follows, on the assumption that $P_0(0) = 1$.

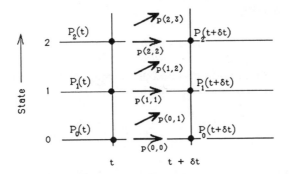

Time $\longrightarrow$

FIGURE 11.2 Schematic relationships among populations $P_n(t)$ and transition probabilities for deriving the equations describing the Poisson distribution of radioactive chain decay. Each node represents a species population at a particular time.

First, consider $P_0(t)$. The probability that the nucleus will be in state 0 at time $t + \delta t$ is equal to the probability that it was in state 0 at time t and did not go to state 1. Thus

$$P_0(t + \delta t) = P_0(t)[1 - p(0 \to 1)], \tag{11.8}$$

which may be rewritten

$$P_0(t + \delta t) - P_0(t) = - P_0(t)p(0 \to 1) \tag{11.9}$$
$$= - P_0(t)\mu \, \delta t.$$

Note that we assume the jump probability, $p(0 \to 1) = \mu \, \delta t$, is not a function of time in this simplified discussion. Dividing by δt and taking the limit of small δt in the usual manner, we find the differential equation

$$\frac{dP_0(t)}{dt} = - P_0(t)\mu. \tag{11.10}$$

The solution is given by equation (11.6).

This method may be extended to the probability of any state, $P_n(t)$, by referring to Figure 11.2. For example, the probability of state 1 at $t + \delta t$ is the probability of state 0 at t followed by a transition $0 \to 1$ in the interval δt plus the probability of state 1 at time t followed by *no* transition in the interval δt.

Symbolically, the process is

$$P_1(t + \delta t) = P_0(t)p(0 \to 1) + P_1(t)p(1 \to 1),$$

which may be rewritten

$$P_1(t + \delta t) = P_0(t)\mu\,\delta t + P_1(t)(1 - \mu\,\delta t). \qquad (11.11)$$

There are no contributions from state 2 in this equation because $p(n + 1 \rightarrow n) = 0$, by assumption. With a little manipulation we find

$$\frac{P_1(t + \delta t) - P_1(t)}{\delta t} = [P_0(t) - P_1(t)]\mu \qquad (11.12)$$

and in the limit $\delta t \rightarrow 0$, this gives the differential equation

$$\frac{dP_1(t)}{dt} = [P_0(t) - P_1(t)]\mu. \qquad (11.13)$$

Thus the change of the probability of state 1 has a positive contribution corresponding to transitions from state 0 and a negative contribution representing transitions out of state 1.

Exercise 11.2
By considering the nth state and paraphrasing the previous argument, show that equation (11.13) can be generalized to

$$\frac{dP_n(t)}{dt} = [P_{n-1}(t) - P_n(t)]\mu.$$

The general equation may be written

$$\frac{dP_n(t)}{dt} + \mu P_n(t) = -\mu P_{n-1}(t). \qquad (11.14)$$

We solve this successively for $n = 0, 1, 2, \ldots$, using $P_{-1}(t) = 0$, because the system cannot be in state -1. At each stage, we assume the right-hand side to be known. Starting with $n = 0$, we have equation (11.10), whose solution is given above, $P(0, t) = \exp(-\mu t)$. Setting $n = 1$, in the general equation, we have

$$\frac{dP_1(t)}{dt} + \mu P_1(t) = \mu P_0(t). \qquad (11.15)$$

We solve this by use of an integrating factor. Multiplying through by $\exp(\mu t)$, using the expression for $P_0(t)$, and rearranging slightly, equation (11.15) becomes

$$\frac{d[P_1(t)\exp(\mu t)]}{dt} = \mu. \qquad (11.16)$$

Multiplying by dt and integrating, we find

$$P_1(t)\exp(\mu t) = \mu t$$

or

$$P_1(t) = \mu t \exp(-\mu t). \tag{11.17}$$

In a similar fashion, each $P_n(t)$ may be derived from the preceding $P_{n-1}(t)$, and we find that the successive probabilities are related by the expression

$$P_n(t) = P_{n-1}(t)\left(\frac{\mu t}{n}\right). \tag{11.18}$$

This expression implies that, for a given value of μt, the probability reaches a maximum where $\mu t = n$ and becomes smaller for larger n. Evaluating successive probabilities we find

$$P_2(t) = \frac{(\mu t)^2 e^{-\mu t}}{2},$$

$$P_3(t) = \frac{(\mu t)^3 e^{-\mu t}}{6},$$

and, by induction,

$$P_n(t) = \frac{(\mu t)^n e^{-\mu t}}{n!}. \tag{11.19}$$

This is the Poisson distribution.

Exercise 11.3
Starting from the expression

$$\sum_{n=0}^{\infty} P_n(t) = 1$$

and the recursion relation

$$P_n(t) = P_{n-1}(t)\left(\frac{\mu t}{n}\right),$$

show that the summation leads to the series expansion for the $\exp(-\mu t)$ and prove by induction the general expression for $P_n(t)$ given above.

Exercise 11.4
Compare the results of your simulation in Exercise 11.1 with the theoretical expectation, $P_n(t)$.

11.3   METEOR ARRIVALS: DIVISIONS OF A TIME-LINE

In this section, we discuss another problem that leads to a Poisson distribution, and we formulate it to stress the relation of the Poisson and Bernoulli distributions.

During nights when there is no meteor shower, a single observer can usually see about 10 "sporadic" (or random) meteors per hour, if the sky is very clear and there is no interference from moonlight. The arrivals of these meteors are independent events, and they can be simulated by randomly placing points on a line, as in Exercise 11.5. When the numbers of meteors, n, that arrived during each hour are tallied, the frequency distribution of n will follow the Poisson distribution. The shape of this distribution will depend only on the mean number, $m \equiv \langle n \rangle$, per hour.

Figure 11.3 shows an example of such data for meteors observed with a radar telescope and divided into 30-second intervals.

Exercise 11.5
Carry out a simulation of meteor arrivals as follows. Consider a "time-line" corresponding to one night's observations, say 10 hours. Randomly assign 100 meteors to times during that interval (so the mean is 10 meteors per hour) and tally the number that arrived during each hour. Construct a frequency table in which $h_n(10)$ is the number of hours in which n meteors appeared. Repeat this about 100 times, or until the shape of the frequency table is well established. Prepare a histogram to display the distribution $h_n(10)$.

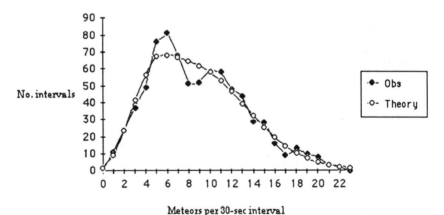

Meteors per 30-sec interval

FIGURE 11.3   Number of intervals in which various numbers of meteors were detected by radar. Data from R. E. McCrosky, *Bull. Astron. Czech.* 8:1, 1957. Observed data are shown with solid dots and theoretical, according to Poisson distribution, are shown with hollow dots. The most common rate during this observational period was about 6 per 30-second interval. These meteors are much fainter and more frequent than naked-eye meteors, which occur at a rate of about 1 per minute on an active night.

We first show that the expected frequency distribution $\langle h_n(10) \rangle$ follows the Bernoulli distribution. To do this, we focus on one particular hour out of the 10, say, the first hour. If a meteor arrives at random during the entire 10-hour interval, it will have the probability $p = \frac{1}{10}$ of arriving during the first hour and $1 - p$ of not arriving during the first hour. The total of 100 meteors make up 100 binary trials, and the probability of n successes (hits in the first hour) is given by the Bernoulli distribution, $B(n, 100)$:

$$B(n, 100) = \left[\frac{100!}{(100 - n)! n!} \right] p^n (1 - p)^{100 - n}.$$

Even for this modest problem, we must evaluate 100!, which often causes an overflow. One way to alleviate this difficulty is to compute the ratio

$$\frac{100!}{(100 - n)!} = 100 \times 99 \times \cdots \times (100 - n + 1),$$

but when n is large, even this calculation can cause trouble. A more powerful method is to develop an approximate relation, by going to the extreme and then simplifying. Suppose, for example, we observe for H hours and count a total of N meteors, letting H and N become very large. The mean rate is $m \equiv N/H$ meteors per hour and is assumed to be constant during the entire observing session. This implies that the probability that a randomly arriving meteor will arrive in a particular hour is $p = 1/H$. The expected frequency $\langle h(n) \rangle$ is given by the Bernoulli distribution (4.2),

$$B(n, N) = \left[\frac{N!}{(N - n)! n!} \right] p^n (1 - p)^{N - n}.$$

What happens to this distribution as we observe meteors for a longer and longer time, so $H \to \infty$, $N \to \infty$, while m remains finite and constant? The result is the Poisson distribution, and we shall now derive this limit.

11.4 DERIVATION OF THE POISSON DISTRIBUTION FROM THE BERNOULLI DISTRIBUTION

Now let us see what form (4.2) takes when p becomes smaller while N becomes larger in such a way that the mean expectation, defined by $m \equiv \langle n \rangle = Np$, remains finite. We first multiply and divide the general expression by N^n and rewrite it as

$$B = \left[\frac{N(N - 1) \cdots (N - n + 1)}{N^n} \right] \left[\frac{1}{n!} \right] (Np)^n (1 - p)^{N - n}.$$

Now we note that, for $N \gg 1$ and $N \gg n$, the first term in brackets is approximately unity, while the last term approximates the exponential as

$$\left[1 - \frac{m}{N}\right]^{N-n} \rightarrow \left[1 - \frac{m}{N}\right]^{N} \rightarrow e^{-m}.$$

Inserting these, we find again the Poisson distribution (11.19):

$$P_n(m) = \left[\frac{m^n}{n!}\right] e^{-m}.$$

The Poisson distribution is a discrete distribution, because n can take only integer values. Furthermore, it is defined by a single parameter, the mean m. Once we know the mean, we can compute the probability for any value. This is different from the Bernoulli distribution for which we had to specify the probability and the number of trials. In deriving the Poisson distribution, we assume that the number of trials is effectively infinite, although the mean number of successes, Np, remains finite. We now return to the arrival times of meteors.

Using the Poisson distribution, we can compute $h_n(10)$ the number of hours in which we are to expect $n = 0, 1, 2, \ldots$ meteors. The probability that n will

TABLE 11.1 Number of Hours in Which n Sporadic Meteors Are Expected (Mean Rate $m = 10$ per Hour; Total of 20 Hours)

Number of Meteors, n	Number of Hours, $h_n(10)$
2	0
3	0.1
4	0.4
5	0.8
6	1.3
7	1.8
8	2.2
9	2.5
10	2.5
11	2.3
12	1.9
13	1.5
14	1.0
15	0.7
16	0.4
17	0.3
18	0

appear during a given hour, if 10 is the mean number during an hour, is

$$P_n(10) = \frac{10^n e^{-10}}{n!},$$

and so if we count for 20 hours we expect the number with n meteors to be $h_n(10) = 20P_n(10)$. These values are given in Table 11.1 (and are plotted in Figure 11.4) for an observer who is assumed to watch the sky for a total of 20 hours.

This table indicates that, out of 20 hours of sky-watching, we would rarely expect to find any hours in which less than 2 or 3 meteors are seen, nor would we expect to see more than 17 or 18 in a single hour. On the other hand, we would expect to see 8 to 12 meteors—roughly the mean rate—in 11 or 12 out of the 20 hours.

Exercise 11.6
Verify that $P_n(10)$ given in Table 11.1 describes the distribution $h_n(10)$ found in the simulation you did in Exercise 11.5.

Such a table can be a guide to the discovery of meteor "showers." (A shower is the result of the Earth's passage through a stream of particles left in the wake of a comet and orbiting about the sun. Each year the Earth passes through many such streams.) For example, if a rate of 15 meteors per hour persists for 3 or 4 hours, this probably is the sign of a shower. The peak rate during the Perseid shower is about 1 per minute and this rate persists for several nights following the 11th of August each year. (Meteor showers appear to come from

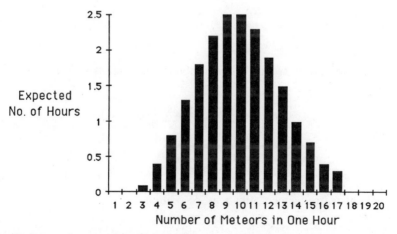

FIGURE 11.4   Expected frequency of hours, $h_n(10)$, with different numbers of meteors when the mean is 10 per hour.

small regions of the sky, called "radiants," and this provides another sensitive test for the presence of a shower.)

Exercise 11.7
Verify that the Poisson distribution approximates the Bernoulli distribution by going back to the horse race in Chapter 4 and recomputing Table 4.2 using the Poisson distribution. First you must evaluate the mean rate m, and then evaluate $P_n(m)$ for various numbers of steps, n.

11.5 MOMENTS OF THE POISSON DISTRIBUTION

Several properties of the Poisson distribution are useful. First, the sum of the probabilities over all possible outcomes n is unity, just as it was for the Bernoulli distribution:

$$\sum_{n=0}^{\infty} P_n(m) = 1.$$

(All the remaining sums in this section are to be understood to be taken over the range $0 < n < \infty$.)

The expectation of the variable n is given by

$$\langle n \rangle = \sum n P_n(m) = \sum \frac{n e^{-m} m^n}{n!}$$

$$= \sum \frac{e^{-m} m m^{n-1}}{(n-1)!}$$

$$= m \sum \frac{e^{-m} n^{n-1}}{(n-1)!} = m. \tag{11.20}$$

The expectation of the second moment about the origin is

$$\langle n^2 \rangle = \sum n^2 P_n(m) = \sum \frac{n e^{-m} m^n}{(n-1)!}$$

$$= m \sum \frac{n e^{-m} m^{n-1}}{(n-1)!}.$$

With a slight rearrangement, this may be written

$$\langle n^2 \rangle = m \sum \left[\frac{(n-1) e^{-m} m^{n-1}}{(n-1)!} + \frac{e^{-m} m^{n-1}}{(n-1)!} \right]. \tag{11.21}$$

The summation of the first term in brackets is just the mean m, while the second term is unity, so we find

$$\langle n^2 \rangle = m(m + 1). \tag{11.22}$$

The second moment about the mean is the variance of the distribution, s^2, and by the shortcut that is developed in Appendix 2, the variance is the mean of the square minus the square of the mean. For the Poisson distribution, this leads to

$$s^2 = \langle n^2 \rangle - \langle n \rangle^2 = m(m + 1) - m^2 = m. \tag{11.23}$$

Thus the variance about the mean equals the mean itself. The dispersion, which is the square root of the variance, equals the square root of the mean. This is a very useful and general rule, as we shall see. In the limit of small p, the Bernoulli distribution obeys the same relationship.

The usefulness of this estimate of the dispersion of the Poisson distribution can be illustrated as follows. Referring back to Table 11.1, we note that the mean rate $m = 10$ meteors per hour implies a dispersion of about 3. This means that we should expect the hourly counts to range typically from $10 - 3 = 7$ to $10 + 3 = 13$ meteors per hour. We can sum the entries in this table and evaluate how many out of the 20 hours will actually fall into this range. The result is $1.8 + 2.2 + 2.5 + 2.5 + 2.3 + 1.9 + 1.5 = 14.7$, so in this case three-quarters of the histogram occurs inside the dispersion.

11.6   APPLICATIONS

(a) "Shot Noise" in the Photons from a Star

During the early days of shot guns, a common way to produce the small lead spheres that make up shot was to let droplets of molten lead fall through the air into water. The droplets would solidify on their descent, and the irregular noise of their impacts was known as "shot noise." This name has been applied to a wide range of phenomena in physics and electrical engineering.

The arrival times of the individual pieces of shot make up a pattern that is mathematically similar to the pattern of meteor appearances. It too is governed by the Poisson law, and the earliest discussion of photon noise appears to have been written in the 18th century, when the corpuscular theory of light (espoused by Isaac Newton, among others) was in vogue. John Michell, an English parish minister with a talent for mathematics and a keen interest in astronomy, suggested that starlight ought to display granularity because the individual photons will fall irregularly on the eye. He suggested this as at least a partial explanation of twinkling.

Michell did not have any way of estimating the number of photons

Time   →

FIGURE 11.5 Starlight is modeled as a train by photons randomly placed on a time-line. The distribution of arrivals in a standard time interval is described by the Poisson distribution.

corresponding to stars of different brightness, so he could not compute the magnitude of this effect, but his prediction was verified during the 20th century, when photon-counting photometry was applied to stars and the shot noise of starlight was found to be an important feature in the light of the fainter stars and the galaxies.

To estimate the shot noise of photons on the human eye when looking at a star near the limit of visibility, we assume that the photons follow the Poisson law (Figure 11.5). In this case, we need only specify the mean rate m at which photons strike the eye from a particular star. More precisely, we need to specify the number of photons that strike the eye in a particular time inverval. The shorter the interval, the smaller will be this number, and the greater will be its variance. The relevant time interval is the flicker-response time of the eye, and it is well known that the eye responds to variations in times as short as about $\frac{1}{20}$ second. (As Michell noted, this interval can be estimated by noting the blurring effect of rapidly moving bright objects. Today, we may use movie images, for example.)

A star of magnitude $V = 5.0$ is comfortably above the limit of visibility and it sheds 4×10^{-8} ergs of light on each square centimeter per second. The pupil of the eye covers approximately 0.05 square centimeters, so 2×10^{-9} ergs enter the eye each second. How many photons does this imply? Each visible photon carries approximately 5×10^{-12} ergs, so we estimate that about 400 photons enter the eye each second. In $\frac{1}{20}$ second, 20 photons would have entered the eye, on the average. Thus $m = 20$, $s = \sqrt{m}$, and for the relative amplitude, we have $s/m = \sqrt{1/m} = 1/4.5$. This implies a fluctuation, which we see as a "twinkling," of about 22 percent in the brightness of the star in intervals of $\frac{1}{20}$ second. (Unfortunately, unless we observe from above the Earth's atmosphere, there will be a large component of twinkling due to the irregularities of the air. We ignore these for this discussion.)

Exercise 11.8

Instead of a $\frac{1}{20}$-second interval, suppose we measure the star's brightness in the interval T. How does the percent fluctuation change with the length of the interval? In particular, how long an interval is required for the shot noise to decrease to 1 percent?

(b) Distribution of Stars on the Sky and Bacteria on a Slide

Just as meteor and photon arrivals can be considered as random events along a time-line divided into equal time intervals, the apparent positions of stars can be considered as random events on the plane of the sky divided into equal areas. In a similar way, the positions of individual bacteria on a slide can be represented by a random arrangement. They obey similar statistical laws.

Deviations from randomness, such as clusters, are not uncommon, and they give important clues to the processes of star formation. A few of the pairings are accidental, but most of them represent physical systems, and one of the most celebrated examples is Beta Capricorni, a bright pair that can be separated with a small binocular under good conditions. The two stars of this pair are 3 arc-minutes or $\frac{1}{20}$ degree apart.

Another of John Michell's contributions to astronomy was to compute the likelihood that such a pair would occur if the stars were scattered haphazardly on the sky. The essence of his calculation is similar to the evaluation of the Bernoulli distribution, and it can be described as follows. Suppose the first star of the pair were located at an arbitrary spot on the sky and compute the chance that another star, if placed randomly, would fall within $\frac{1}{20}$ degree of it. The area of the circle would be about $\pi r^2 = \frac{3}{400}$ square degrees, and this is about $(\frac{3}{400})/34,000 = 2.2 \times 10^{-7}$ of the entire sky. This is the probability that any one star placed at random will fall within $\frac{1}{20}$ degree of an arbitrary star, but it does not finish the calculation. Michell estimated that there were 230 stars on the sky at least as bright as the components of Beta Capricorni. We need the probability that any two of them would be within $\frac{1}{20}$ degree of each other, so we need the number of pairs that could be formed from 230 stars. This is $C(230,2) = 230!/(228!2!) = 26,335$. Multiplying this by the probability for a given pair, we find $p = 0.0058$ for the likelihood of a pair such as Beta Capricorni. (This figure is about one-half the value found by Michell.) So, on this basis, we would guess that the pair is not an accidental arrangement but is probably a system of two stars that had been formed together. Modern measurements show that this is indeed the case. These two stars are traveling on parallel paths through space, and they will remain together for tens of thousands of years. If we were to examine more of the pairs that appear at least this close together we would find that the vast majority are physical pairs, in orbit about each other as the Earth and planets orbit about the Sun. In fact, the majority of all stars are members of groups, some quite large and loose and more difficult to detect.

Another group considered by Michell was the Pleiades, in the constellation of Taurus, and we leave that calculation as an exercise for the reader.

Exercise 11.9
The Pleiades cluster may be described as a group of 6 naked-eye stars in an area of about $\frac{1}{4}$ square degree. Using the facts that there are about 1500 stars as bright as the Pleiades on the entire sky, and that the sky covers about 34,000 square degrees, form your own estimate of the probability that such a group would occur by chance.

Clusters such as the Pleiades are obvious, but a test for randomness in cases that are less obvious can be built on the Poisson distribution. The chart in

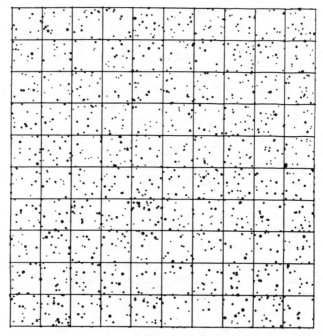

FIGURE 11.6   Chart of a real star field with a 1-degree grid superposed. Stars to about magnitude 9 are shown (about 10 times fainter than naked-eye stars).

Figure 11.6 shows the positions of stars down to about the 9th magnitude in a small region of the sky. The region has been divided into 1-degree squares, and the star may be counted in each square. The numbers of squares with varying numbers of stars may then be compared with the Poisson distribution.

Before applying this test to a real star field, we apply it to an artificial field, in which each star's x and y coordinates are derived from a random-number generator. An example of such an artificial field is shown in Figure 11.7. In the first frame, the stars are plotted without a grid, and the second and third show superposed grids of different spacing. The field contains 100 stars, all of the same magnitude, and the mean number of stars per cell is 6.25 in the coarser grid and 1.56 in the finer grid.

Many of the cells in the finer grid are empty. We ask whether this is consistent with the Poisson distribution, and in Figure 11.8 are plotted the distributions of cell populations for the two grids. The Poisson distributions with the appropriate mean populations are also shown, and the agreement is good for the finer grid. The argeement for the coarser grid is not quite so good, but the discrepancies are characteristic of a randomly generated field, and they roughly follow the rule we have established earlier: the fluctuations are comparable to the square roots of the mean numbers. (This follows from the relationship $s^2 = m$ obeyed by the Poisson distribution.) The coarser grid, having fewer squares, is

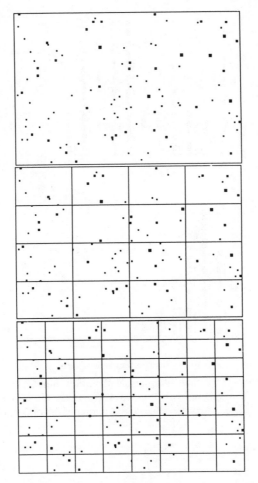

FIGURE 11.7   Artificial star field with grid of 16 and 64 squares superposed.

expected to show larger fluctuations from the theoretically expected Poisson curve.

The histogram for the real star field is shown in Figure 11.9. Again, the discrepancies are comparable to the square roots of the counted numbers, so we conclude that the distribution is essentially random.

Exercise 11.10

(a) With your computer, construct a star field in which the distribution is not entirely random. For example, start with a random field of about 1000 stars, divide it into $S \sim 100$ equal areas giving star counts $n_i(i = 1, \dots, S)$, and then place additional stars into squares that have more than the average number, $\langle n \rangle = \sum n_i/S$. This can be accomplished by multiplying each n_i by a constant factor f and truncating to an integer. Do this for various values of f to see whether you can detect the departure from randomness.

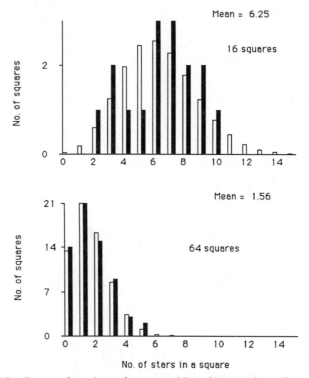

FIGURE 11.8   Counts of numbers of squares with various numbers of stars for artificial star field (Figure 11.7), compared with Poisson distribution. This diagram indicates the typical ranges of deviations between theory and actual counts that are to be expected from small samples.

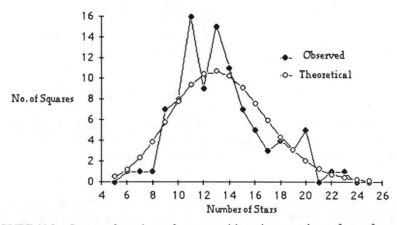

FIGURE 11.9   Counts of numbers of squares with various numbers of stars for a real star field (Figure 11.6), compared with Poisson distribution. The disagreements between theory and observation are in accord with expectation, so we may conclude that the star field is essentially random.

(b) For various f, construct histograms of the frequency distribution of n_i and compare them with the Poisson distribution to see whether you can detect the imposed deviations from randomness. Apply the chi-squared test of Chapter 1 for a quantitative measure of such a deviation. How does the detectable value of f compare with what you found in part (a)?

REFERENCE

Papoulis 1965

12

THE NORMAL DISTRIBUTION

In the previous chapter, while deriving the Poisson distribution from the Bernoulli distribution, we assumed the number of trials to satisfy $N \gg 1$ and we placed the additional restriction that the probability of success in each trial, p, was very small so the expectation of the mean number of successes, Np, remains finite. In this chapter we set $p = \frac{1}{2}$, require $N \gg 1$, and we find the "normal," or gaussian, distribution. We also show how the continuous function describing the normal distribution may be derived from a differential equation.

12.1 THE NEED FOR A FURTHER APPROXIMATION TO THE BERNOULLI DISTRIBUTION

Many examples of the Bernoulli distribution for large N can be conjured up, and they quickly lead us to seek an approximation that is convenient when N becomes very large.

(a) Brownian Motion and the Long Galton Board

In Chapter 3, we discussed the Galton board as an analogy for a random walk in one dimension. Such a walk is also a simplified model of "Brownian motion," the observed irregular motion of small bits of matter in a gas or liquid. Brownian motion is responsible for the diffusion of one gas through another, for example, the aroma of a flower that diffuses through the air. Each molecule of perfume travels a random path from the flower, and they gradually spread into a larger and larger volume.

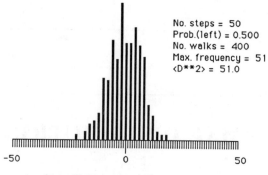

No. steps = 50
Prob.(left) = 0.500
No. walks = 400
Max. frequency = 51
<D**2> = 51.0

-50   0   50

Steps Right or Left of Start

FIGURE 12.1 Frequency of endpoints of 400 simulated walks on Galton board. The histogram becomes smoother as the number of walks increases, and it approaches the bell-shaped "normal" distribution.

Our earlier discussion of the Galton board was limited to a relatively small number of steps, say, $N < 10$. Now we let N become large and plot the simulated outcome of a long Galton board for $N = 50$ in Figure 12.1. The result is a nearly symmetric, bell-shaped curve that is quite smooth. The smoothness arises from the large number of possible outcomes, because the net displacement of such a walk can be anywhere from $+N$ to $-N$ (although these extremes are highly unlikely). The smoothness suggests that we might approximate the discrete set of outcomes with a continuous function, and the purpose of this chapter is to discuss the nature of that function.

(b) Throwing a Large Number of Dice

As another example, suppose we extend Exercise 5.3 and toss a very large number of dice. What will be the shape of the resulting frequency distribution of the sums?

First, suppose a single die of M sides is tossed repeatedly. Let $S_1(M)$ be a random variable representing the points shown by the die. It can take the equally probable values $\{1, 2, \ldots, M\}$. Furthermore, let $\mu_1 \equiv \langle S_1 \rangle$ be the expected mean value of the points for the single die. If the die is not loaded, we expect

$$\mu_1 = \frac{1 + 2 + \cdots + M}{M}. \tag{12.1}$$

Exercise 12.1
Show that (12.1) gives $\mu = 3.5$ for a six-sided die.

Now if we toss N similar dice and sum the points, we find the random variable $S_N(M)$, which can take the values $\{N, N+1, \ldots, MN\}$. We seek the mean and the variance of the sum, $S_N(M)$. In what follows, we assume a dependence on M without explicitly indicating it.

Let μ_N be the expected mean, $\langle S_N \rangle$. We note that each die will, in the mean, contribute μ_1 to the sum, so the expected mean is N times the expected mean for a single die,

$$\mu_N \equiv \langle S_N \rangle = N\mu_1. \tag{12.2}$$

So much for the mean of the sum of a large number of dice. What about its variance? To approach this problem, imagine throwing the dice one by one and computing the running sum. The variance after N dice is defined as

$$s^2 \equiv \sum \frac{(S_N - \langle S_N \rangle)^2}{N}. \tag{12.3}$$

The growth of the sum can be considered as a random walk with a variable step-length. By (5.5), the variance of the endpoints is given by the number of steps times the mean-squared step-length. Let π_i be the point shown by the ith die, $\pi_i = \{1, \ldots, M\}$. To compute the step-length we need the deviation from the mean at each step, so the step for the ith die is

$$\sigma_i = \pi_i - \mu_1. \tag{12.4}$$

The expected variance is $N\langle \sigma_i^2 \rangle$, and this gives the following expression for the expected variance:

$$\langle s^2 \rangle = N\langle (S_1 - \mu_1)^2 \rangle. \tag{12.5}$$

Exercise 12.2
Show that the mean-squared step-length for a six-sided die ($M = 6$) is

$$\langle (S_1 - \mu_1)^2 \rangle = \frac{1}{M}[(1 - 3.5)^2 + (2 - 3.5)^2 + (3 - 3.5)^2 + \cdots] = 2.92,$$

so (12.5) gives $\langle s^2 \rangle = 1.71N$ for the variance of the sum of N six-sided dice.

Exercise 12.3
Carry out a simulation of the throwing of $N = 200$ six-sided dice. Compute the sum of the points, S_{200}, and repeat this process a large number of times, until the shape of the frequency distribution of sums becomes apparent. Construct a histogram in which $F(S_{200})$ is plotted against S_{200}. Evaluate the mean and the variance of $F(S_{200})$. How closely do they obey the theoretical relations for the expected values and the variance, $\mu_{200} = 200\langle S_1 \rangle = 700$, $s^2 = 584$?

12.2   DERIVATION OF THE NORMAL DISTRIBUTION FROM THE BERNOULLI DISTRIBUTION

Consider a long Galton board whose outcomes are described by (4.2):

$$B(N, m) = \left[\frac{N!}{[(N + m)/2]! [(N - m)/2]!} \right] p^n (1 - p)^{N-n},$$

where N is the number of steps and m is the net displacement. For a symmetric walk, $p = \frac{1}{2}$, we have

$$B(N, m) = \left[\frac{N!}{[(N + m)/2]! [(N - m)/2]!} \right] \left(\frac{1}{2} \right)^N. \tag{12.6}$$

We are interested in the limit $N \to \infty$ and will confine our attention to the region near the origin, $m \ll N$. Equation (12.4) may be reduced using Stirling's formula:

$$\ln n! \approx (n + \tfrac{1}{2}) \ln n - n + \tfrac{1}{2} \ln 2\pi \quad (\text{as } n \to \infty).$$

It can easily be shown that

$$\ln B(N, m) \approx (N + \tfrac{1}{2}) \ln N - \tfrac{1}{2}(N + m + 1) \ln [N(1 + m/N)/2]$$
$$- \tfrac{1}{2}(N - m + 1) \ln [N(1 - m/N)/2]$$
$$- \tfrac{1}{2} \ln 2\pi - N \ln 2. \tag{12.7}$$

And as $m \ll N$, we may use the series expansion

$$\ln (1 + \varepsilon) \approx \varepsilon - \frac{\varepsilon^2}{2N^2} \tag{12.8}$$

to find

$$\ln B(N, m) \approx -\frac{1}{2} \ln N + \ln 2 - \frac{1}{2} \ln 2\pi - \frac{m^2}{2N},$$

which is equivalent to

$$B(N, m) \approx \sqrt{\frac{2}{\pi N}} \exp \left(\frac{-m^2}{2N} \right). \tag{12.9}$$

This is the normal or gaussian distribution.

Exercise 12.4
Supply the missing steps in the preceding derivation, and verify the accuracy of the approximation for $N = 10$.

12.3   SIMULATING THE NORMAL DISTRIBUTION

The nature of the normal distribution can be seen by simulating with a simple computer program. Such a program, which we will call "Gauss," is listed below in BASIC. (Program adapted from Miller, 1981.)

Listing of Basic Program Gauss

```
Rem  Produces Gaussian Random Number, G, With Mean D1, Variance D2.
Randomize Rem Initialize Basic's Random Number Generator
Read D1,D2 Rem  Desired Mean And Variance In Data Statement below
S = 0
N = 12
For I = 1 To N
                Y = Rnd - 0.5 Rem  Random Number In Range -0.5 >Y >0.5
                S = S + Y  Rem  Adds New Random Number To Sum
Next I
G = S*D2 + D1 Rem  New Variable With Desired Mean And Variance
Data 0,1  Rem  Input Values of D1 And D2
End
```

The program asks for a random seed (**Randomize**), reads two parameters from a data statement, and initializes the variable $S = 0$. It then calls a series of N random numbers (called by **Rnd**) that are uniformly distributed in the range 0 to 1.0, and transforms to Y. The sum of the Ys is a normally distributed random number with zero mean and unit variance. This is then transformed to the new variable, G. The frequency with which various values of G are generated will imitate a normal distribution with arbitrary mean $D1$ and variance $D2$.

To see that the spread of G given by the program Gauss will be $D2$ as long as $N = 12$, recall the earlier discussion of the random walk. If a walk proceeds with backward or forward steps of length Y, the walker will, after N steps, have a mean displacement $\langle Y \rangle = 0$, and a mean-squared displacement (variance) given by $\langle Y^2 \rangle = NY^2$.

If the steps have varying length, we merely replace Y^2 by the mean-squared step-length. Suppose the fraction $p(Y)dY$ of the steps had lengths in the interval Y to $Y + dY$. We can compute the mean length and the mean-squared length of the step from the integral

$$\langle Y^2 \rangle = \int_0^{1/2} p(Y)Y^2 dx = \tfrac{1}{12}. \tag{12.10}$$

In the present case $p(Y) = 1$, and these relations lead to variance of Y equal to $N\langle Y^2 \rangle = 1$, and variance of G equal to $D2$.

Exercise 12.5
Prepare and run a program to compute 100 numbers using Gauss as a subroutine. Prepare histograms of the output using different values of N and $D2$ and verify the behavior of G.

12.4 DIFFERENTIAL EQUATION DESCRIBING DIFFUSION

By considering a finely divided region of space, we may set up a differential equation for the density of particles undergoing random walks. This is a diffusion process, and we may follow the probable locations of one species of particles through a lattice provided by another species.

Divide a one-dimensional region into numbered cells ($i = 1, 2, 3, \ldots$) of width δx, and for simplicity, assume that each particle is located at the center of its cell. The cells may be imagined to shrink to small regions separated by distance δx. This is shown schematically in Figure 12.2, where a vertical row of cells represents the array of cells stretched along a line. Time increases to the right in equal increments δt. At each time-step, all the particles move randomly up or down one cell with equal probabilities, $\frac{1}{2}$. (This is equivalent to the Galton board tipped on its side.) Thus, for each particle, the transition rule at each time step is $j \to j + 1$ and $i \to i \pm 1$.

Let $n(i, j)$ be the expected number of particles in cell i at time j. We seek an expression for $n(i, j + 2)$ in terms of $n(i, j)$ and we proceed as follows. First, we express the expected number $n(i, j + 2)$ in terms of the particle densities in the previous time. This is just the number of entries, because all the original particles have departed:

$$n(i, j + 2) = \tfrac{1}{2}n(i + 1, j + 1) + \tfrac{1}{2}n(i - 1, j + 1). \tag{12.11}$$

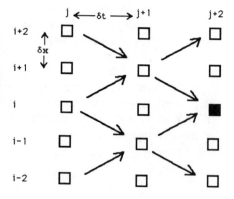

FIGURE 12.2 Schematic diagram of diffusion of particles in a one-dimensional space. Distance is vertical and time is horizontal. All particles are assigned to small regions separated by distance δx, and at every time-step δt, they jump to an adjacent cell, as indicated by the arrows. A differential equation describing the smoothed particle densities is derived by finding an expression for $n(i, j + 2)$ from $n(i, j)$ and letting the spacing go to 0.

Second, we write each density on the right-hand side in terms of the densities in row j, as follows:

$$n(i+1, j+1) = \tfrac{1}{2}n(i+2, j) + \tfrac{1}{2}n(i, j), \tag{12.12}$$

$$n(i-1, j+1) = \tfrac{1}{2}n(i, j) + \tfrac{1}{2}n(i-2, j). \tag{12.13}$$

Inserting these into the previous expression, we have

$$4n(i, j+2) = n(i+2, j) + 2n(i, j) + n(i-2, j),$$

and substracting $4n(i, j)$ from both sides, we have for the change in the density in cell i

$$4[n(i, j+2) - n(i, j)] = [n(i+2, j) - n(i, j)] - [n(i, j) - n(i-2, j)]. \tag{12.14}$$

The left-hand side is four times the change in cell i during two time-steps; it equals the difference of the two bracketed terms.

Readers who are not familiar with calculus may wish to skip from here to the start of the next section and consider this as merely another way of describing the set of random walks on a Galton board.

But let us push on. Each of the bracketed terms may be written in terms of a slope in a particular direction (a *partial derivative*); so we find

$$4\delta t \left[\frac{\partial n(i, j)}{\partial t} \right] = \delta x \left[\frac{\partial n(i+1, j)}{\partial x} - \frac{\partial n(i-1, j)}{\partial x} \right]. \tag{12.15}$$

The term $\partial n(i, j)/\partial t$ is the rate at which the density $n(i, j)$ changes in time. This rate equals the difference between the inflow and outflow to the cell, represented by the bracketed term on the right.

Next, we write this bracketed term on the right as a second derivative,

$$\frac{\partial n(i+1, j)}{\partial x} - \frac{\partial n(i-1, j)}{\partial x} = 2\delta x \left[\frac{\partial^2 n(i, j)}{\partial x^2} \right],$$

and when we insert this into the previous expression, we find the differential equation describing diffusion,

$$\frac{\partial n(i, j)}{\partial t} = D \frac{\partial^2 n(i, j)}{\partial x^2}, \tag{12.16}$$

where we have defined the "diffusion coefficient,"

$$D \equiv \frac{\delta x^2}{2\delta t}. \tag{12.17}$$

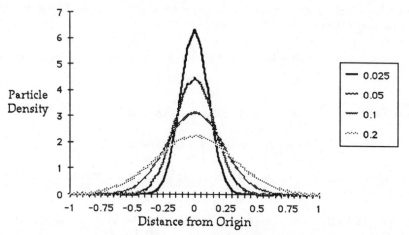

FIGURE 12.3   Solution to the diffusion equation for $n(x, t)$ at various values of t shown in the legend. The area under the curves is the total number of particles, and it remains constant while the peak value drops.

The following function satisfies this equation, as may be verified by carrying out the differentiation:

$$n(x, t) = \frac{N_0}{2\sqrt{\pi Dt}} \exp\left(-\frac{x^2}{4Dt}\right), \tag{12.18}$$

where the coefficient is chosen so the integral over space gives the constant total number of particles, $\int n(x, t)dx = N_0$.

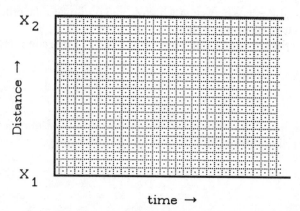

FIGURE 12.4   Space–time region in which one-dimensional diffusion takes place. In order to specify the solution completely, we must specify the density $n(x, 0)$ along the $t = 0$ axis at the left as well as the values of $n(x, t)$ at $x = X_1$ and $x = X_2$ at the ends of the region. The differential equation then determines the densities in the interior of the region.

This is the "normal" curve, and it describes diffusion from a point. Thus we recover our earlier result that the random walk from a point approaches the normal curve in the limit of a large number of steps. As time increases, this curve becomes lower and wider, although its area remains constant, because no particles are lost (Figure 12.3).

To find the solution for a particular problem, we must also specify the starting conditions, $n(x, 0)$, at $t = 0$, and we must specify the boundary conditions satisfied by $n(X_1, t)$ and $n(X_2, t)$, at the limits of the system, as indicated in Figure 12.4. If the boundaries are absorbing, we set $n(X, t) = 0$, and if they are reflecting, we set the slopes to zero, $\partial n(X, t)/\partial x = 0$.

A detailed discussion would take us beyond the scope of this book, but we note that the speed of the diffusion process increases with the diffusion coefficient D, which increases with $\delta x^2/\delta t$. If the time between particle jumps, δt, is decreased, the diffusion becomes more rapid, and the same occurs if δx, the distance jumped, increases. This implies that diffusion is more rapid in a hotter gas, as well as in a gas of lower density, in which particles can jump a greater distance between collisions.

REFERENCES

Hersch and Griego 1969
Miller 1981

PART THREE

DYNAMIC SIMULATIONS

13

GAS OF ATOMS NEAR EQUILIBRIUM

Imagine a gas of atoms confined by opaque insulating walls. This is called a *closed system* because its atoms and total energy are trapped by the impervious walls. Let us try to picture the scene in microscopic detail. One atom bounces off the wall, flies momentarily in a straight line and strikes another atom, giving up some of its energy of motion and rebounding lethargically to one side, where it strikes another atom and picks up speed. In each collision, the atom gains or loses energy and flies off in another direction. Part of the time, it moves faster than most of its neighbors, then slower. If we watch it long enough, we find that, on the average, it has the same energy as its neighbors. No atom is energetically privileged in this chaotic society.

The total energy of the gas remains constant, trapped by the walls, but at any instant it is not divided equally among the atoms. For a brief moment, some have more than their share and others have less. Then the tables turn and the paupers become the wealthy. At any instant, we may count the numbers of atoms with different amounts of energy and represent the result by a histogram, called the *energy distribution*.

It is a remarkable feature of *randomly shuffled closed systems* that the shape of the energy distribution does not depend on the details of the shuffling process. That is, the average partitioning of energy among atoms is not affected by the details of the process by which the energy is transferred from one atom to another. This chapter will explore these properties by considering a simple gas of neutral atoms. In Chapter 14 we will consider a mixture of atoms and photons.

13.1   RANDOM SHARING OF ENERGY

The total energy associated with the atoms of a gas can appear in three forms:

(1) *Energy of Motion*: kinetic energy, which is computed from the atom's mass and speed with the expression $E = mv^2/2$.

(2) *Internal Energy*: excitation corresponding to electrical interaction among the nucleus and the electrons in an atom.

(3) *Potential Energy*: energy of position, if the mass of the atom is subject to a gravitational field or if it is electrically charged (ionized) and feels an electric field.

There are many ways in which we may follow the history of such a gas. In Chapter 7 we described a kinetic method. Rather than following the trajectories of the atoms to find which pairs collide during any interval, we arbitrarily select a pair of particles, compute their energy, and divide it randomly between them. Having carried out this "collision," we then select another pair at random and repeat the process.

In order to represent the results of this simulation, we divide the range of kinetic energy into equal steps and count the numbers of atoms in each energy step, forming a histogram. The result of the kinetic simulation is independent of the details of the shuffling process.

13.2   ELASTIC SHARING OF ENERGY: A DYNAMIC MODEL

We now turn to a "dynamic" model based on Newton's laws. We assume, as before, that all collisions conserve the total kinetic energy—they are elastic. Also, the atoms are confined to motions in a plane, rather than being free to move in three-dimensional space. This assumption affects the manner of counting energy cells, as we will see.

Suppose there are two types of atoms, heavy and light. The horizontal and vertical speeds of the atoms will be designated u and v, respectively. Histograms of desired quantities (atom heights, speeds, kinetic energies) are computed after every 20 or so time steps and are added to previous histograms to produce the time-averaged histograms.

We will now describe the physics and the programming of this simulation.

(a)   Underlying Physics

(1) The atoms all start with the same speed from random positions and in random directions in a two-dimensional space. They move with constant velocity until they collide with the wall or with another atom.

(2) Collision with a vertical side wall leads to reversal of the horizontal

component of motion, $u \to -u$, while collision with the top or bottom leads to reversal of the vertical motions, $v \to -v$.

(3) The atoms are given a finite size, r, and when the centers of two atoms are closer than the distance $2r$, they are said to have collided. The orientation of the line between the centers of the atoms is found from the positions of the atoms, and the effect of the collision is then computed as though the atoms were smooth, elastic billiard balls.

(b) Computational Procedure at Each Time-Step

To reduce the number of arithmetic operations, the time interval is taken to be unity. (To speed the simulation, the time-consuming calculations ought to be done in fixed-point arithmetic, unless a floating-point processor is available. In some computing languages it is convenient to assign each atom a rectangular region of two-dimensional space. The location of the atom is then equated to the location of the upper left-hand corner of the rectangle.)

(1) At each time-step, all atoms are advanced to their new positions with the relations: $x \to x + x$vel, $y \to y + y$vel.

(2) The sides of each atom are compared to the positions of the corresponding walls. If, for example, the left side of an atom is to the left of the left wall, its horizontal speed is reversed and the atom is offset to a position just inside the wall. Note that an atom can collide with two walls in the same time-step (if it comes close enough to the corner of the box). Hence its position with respect to all walls must be checked at each time step.

(3) To detect collisions between pairs of atoms, the location of each atom (corner of its rectangle) is compared with the interior region of every other atom. (We assume that three or more atoms do not collide at the same time.) The handling of each collision is based on the assumption that the atoms behave as slippery disks. Their velocities at the time of contact are decomposed into components along and perpendicular to the line of contact, indicated by heavy and light arrows in Figure 13.1.

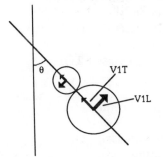

FIGURE 13.1 At the instance of collision between two disklike particles, the velocities are decomposed into components along (V1T) and perpendicular to (V1L) the line of contact. The angle θ indicates the orientation of the line of contact. (See program listing.)

The components of velocity perpendicular to the line of contact are not changed by the collision, because there is no impulse in that direction if the disks are "slippery." Thus we need only evaluate the new speeds along the line of centers, as though we were solving a head-on collision. To do this, we first transform to a coordinate system moving with the center of mass and then impose conservation of momentum. After the new components along the line of centers have been found, we transform back to the laboratory system. The following is a procedure (adapted from a version kindly provided by Professor Edward Purcell) to compute the final velocities for atoms i and j. The conservation of energy is not explicitly used in this calculation, so it can provide a check on the validity of the results. This procedure is invoked when a collision between atoms is detected.

```
Procedure PairHit;  {i,j: integer}
{Algorithm from E. Purcell for collision of slippery disks. }
{The speed across line of centers at instant of contact is unchanged}
{we need only solve conservation of momentum along line of centers.}
{Energy conservation will take care of itself if calculation is correct.}
{This is written in fixed point arithmetic: fixmul(a,b) = a*b, etc.}
   var
         costh, sinth, dd, w, r, dx, dy, v1L, v2L, v1T, v2T, summass: fixed;
   Begin
   summass := atom[i].mass + atom[j].mass;  {auxiliary}
   dx := atom[j].x - atom[i].x;  {x-component of particle separation}
   dy := atom[j].y - atom[i].y;  {y-component of particle separation}
   r := fsqrt(fixmul(dx, dx) + fixmul(dy, dy));  {separation of centers}
   costh := fixdiv(dy, r);{costh = cos(theta),
{theta is angle of line of centers at contact}
   sinth := -fixdiv(dx, r);
{Compute speed of atoms across line of centers}
   v1L := fixmul(atom[i].hvel, costh) + fixmul(atom[i].vvel, sinth);
   v2L := fixmul(atom[j].hvel, costh) + fixmul(atom[j].vvel, sinth);
{Compute speed of atoms along line of centers}
   v1T := fixmul(atom[i].vvel, costh) - fixmul(atom[i].hvel, sinth);
   v2T := fixmul(atom[j].vvel, costh) - fixmul(atom[j].hvel, sinth);
{Find w, velocity of center of mass along line of centers in 2 steps}
{Step 1}
   w := (fixmul(atom[i].mass, v1T) + fixmul(atom[j].mass, v2T));
{Step 2}
   w := fixdiv(w, summass);
   dd := v2T - v1T;  {speed of approach along line of centers}
{Compute new speeds along line of centers}
   v1T := w + fixdiv(fixmul(dd, atom[j].mass), summass);
   v2T := w - fixdiv(fixmul(dd, atom[i].mass), summass);
```

{Now transform back to lab system}
    atom[i].hvel := fixmul(v1L, costh) - fixmul(v1T, sinth);
    atom[j].hvel := fixmul(v2L, costh) - fixmul(v2T, sinth);
    atom[i].vvel := fixmul(v1T, costh) + fixmul(v1L, sinth);
    atom[j].vvel := fixmul(v2T, costh) + fixmul(v2L, sinth);
{Could compare original and final kinetic energies to verify calculation}
    End;

The simulation is initiated by specifying the numbers of atoms of each mass and specifying the speed and location of each atom. Then the positions are stepped forward in time, assuming no forces on the atoms between collisions. After each time-step, the colliding atoms are identified and the procedure **PairHit** is called with the parameters for the colliding atoms. The histogram of speeds and energies is computed at regular intervals and averaged.

Figures 13.2a shows the averaged energy histogram for a long simulation in which there are two types of atom, having masses of 1 and 4. The histogram is plotted separately for each type of particle, and we see that the energy distributions are nearly identical, and this is a reflection of the equipartition of energy. The heavier particles, consequently have smaller speeds. The smallest energy is the most common, as is characteristic of the Boltzmann distribution.

Figure 13.2b show a histogram of speeds for the same simulation. Here we note that the most common speed is not the smallest but is an intermediate value.

The difference between the shapes of the energy and speed histograms in Figure 13.2a and 13.2b is explained by referring to Figure 13.3 which displays

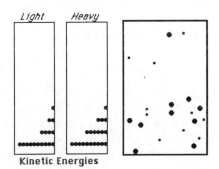

FIGURE 13.2a   Snapshot of a gas of two types of atoms colliding in a box, shown on the right. The large and small dots represent heavy ($m = 4$) and light ($m = 1$) atoms, respectively. Histograms of the time-averaged kinetic energy distributions are shown on the left, with energy increasing upward. Atoms with low energy are the most common, and the heavy and light atoms have the same mean energy, in spite of their difference in mass. This is a universal property of equilibrium gases, known as "equipartition" of energy.

FIGURE 13.2b Similar to Figure 13.2a, except that the speed distributions of the heavy ($m = 4$) and light ($m = 1$) atoms are plotted. Speed increases upward and we see that the most prevalent speed of the light atoms is approximately twice that of the heavy atoms.

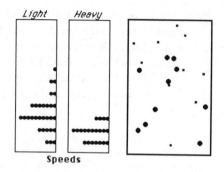

FIGURE 13.3 In this momentum space for two-dimensional gas each particle is represented by the x-component and y-component of its momentum. The number of available cells in each range $p-p+dp$ is proportional to the area $2\pi p\, dp$. The corresponding kinetic energy is $E = p^2/(2m)$. According to the uncertainty principle, the momentum space is divided into cells of area $\tau \equiv dp_x\, dp_y = N\hbar^3/V$, where N is the number of atoms and V is the geometric volume of the gas.

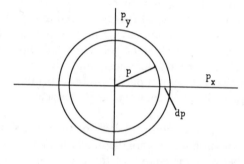

the momentum space for a two-dimensional gas in which motions are confined to the x and y directions. We need first to find the number of cells g_E available in each energy interval (see Chapter 7). The atoms in the circular ring have momentum p corresponding to kinetic energy $E = p^2/(2m)$. The area of the ring is $2\pi p\, dp$ and the number of cells is $g_E\, dE = 2\pi p\, dp/\tau = 2\pi m\, dE/\tau$.

This implies that $g_E = $ constant, so the distribution of energies in a two-dimensional gas is

$$N(E)\, dE = g_E n_E\, dE \sim \exp\left(\frac{-E}{kT}\right) dE. \tag{13.1}$$

The distribution of speeds v may be found by equating the numbers in related intervals: $N(v)\,dv = N(E)\,dE$. Using $E = mv^2/2$ and $dE = mv\,dv$ we find

$$N(v)\,dv = mvN(E)\,dv \sim v\exp\left(\frac{-mv^2}{2kT}\right)dv. \qquad (13.2)$$

The function $N(v)$ rises from 0 at $v = 0$ and then falls off, as in Figure 13.2b.

(c) Comment on Detailed Balancing

Suppose we follow the history of a single particle. Specifically let us plot the x-component of speed as a function of time. In short intervals, this speed fluctuates widely, but if we take long-term averages we will find much the same value in each long interval. This implies that *the global effect of the increases of speed are balanced by the decreases*. The net effect is null. This is the weak form of the principle of detailed balance for a closed system with stationary walls. The same applies to any particle in the gas, if we hold the volume constant and let the gas come to equilibrium.

A stronger form of the principle can be demonstrated if we keep a somewhat closer accounting of the speeds. During a long interval, let us make a histogram of the number of speed *increases in each range*, and let us make a separate histogram of the speed *decreases in each range*. If the system is in equilibrium, these two histograms will have identical shapes, as illustrated in Figure 13.4. That is, there will be as many increases by δv as there are decreases by $-\delta v$. This is a stronger form of the principle of detailed balancing.

There is a simple analogy with a bank account which can illustrate the distinction. If my average balance stays the same from month to month, then I know that the total withdrawals each month equal the total deposits. This is the weak form of the principle of detailed balancing. If my withdrawals in *each denomination* ($1 bills, $10 bills, etc.) are balanced by the deposits in those

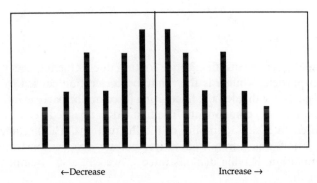

←Decrease   Increase →

FIGURE 13.4   Histogram of number of energy increases and decreases in each energy range. Its symmetry epitomizes the principle of detailed balancing, in which each increase is balanced by an equivalent decrease.

denominations, then my account can be said to satisfy the stronger form of detailed balancing.

There is yet a stronger form, but to see it we must consider different atoms in the gas, and different savings accounts in the bank. By keeping track of all the increases of speed for all atoms, we can demonstrate that the increases in each range of speed $|\delta v|$ are balanced by the decreases. In much the same way, if we look at the activity in all bank accounts and find that, *in each denomination*, the dollar amounts of deposits balance the withdrawals for all accounts taken together, we could say that the flow of money satisfies the strongest form of detailed balancing. This result would be very surprising in banking but it is the rule of the day in an equilibrium gas of atoms.

We may state the strongest form of the principle of detailed balancing in more general terms as follows: *In an equilibrium closed system, the rate of every process is, on the average, balanced by the rate of the inverse process.* There is no elementary proof of this principle, but it can be verified for any physical system. Its importance for us is that it can often help formulate a simulation and it can provide a test for the correctness of a simulation. If a simulation of an equilibrium system does not obey this strongest form of the principle of detailed balancing, then there is an error in the simulation.

(d) Demonstration of the Langevin Equation Describing Brownian Motion with Friction

In Section 5.8, we described the random walk in velocity space by equation (5.14), known as the Langevin equation:

$$\frac{d^2 x}{dt^2} = -\alpha \frac{dx}{dt} + e.$$

In terms of the velocity, $v = dx/dt$, we may express this as

$$\frac{dv}{dt} = -\alpha v + e. \tag{13.3}$$

The coefficient α represents an effective friction due to the impacts of atoms on a Brownian particle, while e is the random acceleration caused by individual impacts. In effect, we divided the fluctuating force on the particle into stochastic (e) and smooth ($-\alpha v$) components. The first led to a jagged velocity curve, and the second led to a slow damping of the fluctuations. The argument for this type of equation was purely qualitative, but we can demonstrate it with a detailed simulation. Having demonstrated it, we will also attempt to show its physical origin.

Instead of considering a mixture of two types of atom, as in the previous section, we consider a box containing identical atoms into which a single heavy particle has been inserted. This is the Brownian particle whose behavior we

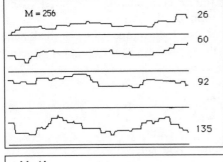

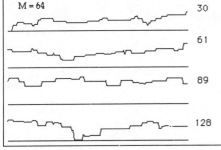

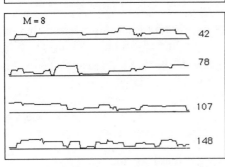

FIGURE 13.5 Momentum versus time for Brownian particles impelled by a gas of atoms. Each set of four segments corresponds to a Brownian particle with the indicated mass in units of the atomic mass. The cumulative number of collisions with atoms is indicated at the end of each strip, and they are approximately the same for all particles. Individual collisions are indicated by jagged steps in the curve. Comparing one mass with another we note that the changes of momentum (speed × mass) accumulate more slowly for the heavy particles. Also, the heavy particles have a larger average momentum, as a result of the tendency toward equipartition of energy with the atoms. These curves are to be compared with similar curves derived by integrating the Langevin equation in Chapter 5.

wish to track. It is defined by its mass and its radius, both measured in terms of the atoms' mass and radius. As before, the atoms collide with each other, in addition to hitting the Brownian particle, and we expect the long-term average energy of the particle to be the same as that of the atoms. Thus, if we start the particle at rest, we expect it to be gradually accelerated by the random impacts, and if we start it with a large speed, we expect it to slow down until its energy fluctuates about the mean energy of the atoms. The rapidity of response to the influence of the atoms will depend on the relative mass of the particle. We demonstrate this behavior in Figure 13.5.

In order to develop an insight into the behavior of the Brownian particle as illustrated in Figure 13.5 and to see the basis for the Langevin equation, we now develop a theory for the dynamics of the particle based on two simplifying assumptions.

Assumption 1 The Brownian particle (index 1) is massive compared to the atoms (index 2), and we describe the ratio by the parameter $\varepsilon \equiv m_2/m_1 \ll 1$. According to the principle of detailed balance, the average values of the kinetic energies of both types of particle are the same, $m_1 \langle v_1^2 \rangle = m_2 \langle v_2^2 \rangle$, and this implies that the average velocities obey

$$\frac{\langle v_1^2 \rangle}{\langle v_2^2 \rangle} = \frac{m_2}{m_1} \equiv \varepsilon, \tag{13.4a}$$

and from this we may infer the approximate relations

$$\frac{\langle v_1 \rangle}{\langle v_2 \rangle} \approx \varepsilon^{1/2}, \tag{13.4b}$$

$$\frac{\langle m_1 v_1 \rangle}{\langle m_2 v_2 \rangle} \approx \varepsilon^{-1/2}. \tag{13.4c}$$

The Brownian particle is, on the average, much slower than the atoms, but by (13.4c) it has a larger momentum.

Assumption 2 All collisions are head-on (Figure 13.6), so we can ignore components of velocity perpendicular to the line of contact. Setting $\theta = 0$ in the equations given in the program listing, we have V1T $= v_1$ and V2T $= v_2$, and for the center of mass speed along the line of centers

$$w = \frac{m_1 v_1 + m_2 v_2}{m_1 + m_2}. \tag{13.5}$$

For the speeds after collision (indicated by primes) we find

$$(m_1 + m_2)v_1' = (m_1 - m_2)v_1 + 2m_2 v_2, \tag{13.6a}$$

$$(m_1 + m_2)v_2' = 2m_1 v_1 - (m_1 - m_2)v_2. \tag{13.6b}$$

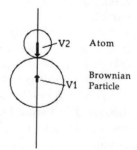

FIGURE 13.6 Geometry of head-on collision between disklike atom and massive Brownian particle. For simplicity, we assume there is no component of velocity across the line of contact. V1 is the velocity of the Brownian particle and V2 ($\gg$V1) is the atom speed. The ratio of masses is $\varepsilon \equiv m_2/m_1 \ll 1$.

If we divide each by m_1 and insert the definition of ε, we find

$$(1 + \varepsilon)v'_1 = (1 - \varepsilon)v_1 + 2\varepsilon v_2, \tag{13.7a}$$

$$(1 + \varepsilon)v'_2 = 2v_1 - (1 - \varepsilon)v_2. \tag{13.7b}$$

Equations (13.7a) and (13.7b) may be simplified. Using Assumption 1 and ignoring terms higher than first order in ε, we find

$$v'_1 - v_1 = 2\varepsilon v_2, \tag{13.8a}$$

$$v'_2 + v_2 = 2v_1. \tag{13.8b}$$

These contain the clues we need. Let us look at (13.8b) first. It tells us what happens to the speed of an atom when it collides with the Brownian particle. Specifically, it says that $v'_2 = -v_2 + 2v_1$, and (13.4b) implies that $v_1 \ll v_2$, for a typical collision, so the atom's velocity is reversed and slightly altered (by twice the speed of the Brownian particle) on each collision. As we shall see in the next section, this behavior may be described by saying that the atom bounces off the Brownian particle as though it were a moving wall.

And from (13.8a) we see that the change of the Brownian particle's speed in a collision is $2\varepsilon v_2$. From (13.4b) we can rewrite the right-hand side for a typical collision as

$$(v'_1 - v_1) = 2\varepsilon \langle v_2 \rangle \approx 2 \langle v_1 \rangle \varepsilon^{1/2} \ll v_1, \tag{13.9}$$

so the change of v_1 is much smaller than v_1 itself. And what about the sign of the change? suppose $v_1 > 0$, so the particle is moving to the right. If it is hit from in front, $v_2 < 0$, and the particle is slowed down. If it is hit from behind, $v_2 > 0$, and the particle is sped up. The same applies if the particle is moving to the left, so we have the following general rule: *The Brownian particle is always slowed down when it is hit from the direction in which it is moving—that is, from the front side.*

Just as a runner in the rain will become wetter on the front side, the Brownian particle will experience more collisions on the forward side. Hence we expect the particle to be gradually slowed down by the net effect of collisions with atoms. This accounts for the resistance term in the Langevin equation (13.3). In order to estimate the rate of slowing down, we need to estimate the difference between the rates of collisions from the front and the back. Referring to Figure 13.7, we see that in a short time δt all atoms in the cylinder $\sigma(v_2 + v_1)\delta t$ strike from the front, while atoms in the cylinder $\sigma(v_2 - v_1)\delta t$ strike from behind, where σ is the cross-sectional area of the Brownian particle and the atoms are assumed to be much smaller. If there are $n/2$ atoms per unit volume moving to the right and to the left, then the *net* rate of collisions from in front is the

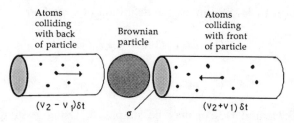

Atoms
colliding
with back   Brownian
of particle   particle

Atoms
colliding
with front
of particle

$(v_2 - v_1)\delta t$ σ $(v_2 + v_1)\delta t$

FIGURE 13.7 Diagram of Brownian particle moving with speed v_1 to right encountering small atoms moving with speed v_2 to right and left. Atoms moving to the left strike the front of the particle; atoms moving to the right strike its back. The numbers striking from each side are given by the number of atoms in the cylinder whose cross-section is σ and whose length is δt times the relative speed of the atoms. Consequently, more atoms strike the front side.

difference of these divided by the time interval, or

$$\text{Net rate of collisions} = n\sigma v_1, \quad \text{from front.} \tag{13.10}$$

Multiplying this rate by the change in momentum per collision gives the effective force of resistance felt by the Brownian particle:

$$\text{Force felt by particle} = -n\sigma v_1 2m_1 \varepsilon v_2. \tag{13.11}$$

According to the model provided by the Langevin equation (multiplied by m_1), we are to represent this force by a term of the form

$$\alpha m_1 v_1 = n\sigma v_1 2m_1 \varepsilon v_2$$

and from this we infer that

$$\alpha = \frac{2n\sigma m_2 v_2}{m_1}, \tag{13.12}$$

after using the definition of ε.

What does this imply about the rate at which the particle's speed is damped? If we ignore the stochastic term e in (13.3), we may integrate at once to find

$$v_1(t) = v_1(0)\exp(-\alpha t). \tag{13.13}$$

The corresponding damping time is given by

$$\tau_{\text{damp}} \equiv \frac{1}{\alpha} = \frac{m_1}{2n\sigma m_2 v_2}. \tag{13.14a}$$

We may also define the collision time, τ_{coll}, as the time between collisions with atoms. It is the reciprocal of the collision rate, so

$$\tau_{coll} = \frac{1}{2n\sigma v_2} \tag{13.14b}$$

and this provides an instructive expression for the damping time,

$$\tau_{damp} = \frac{\tau_{coll} m_1}{m_2} = \frac{\tau_{coll}}{\varepsilon}. \tag{13.15}$$

This states that the damping of the particle's speed will take place relatively slowly—on a time scale that is a factor $1/\varepsilon$ longer than the time between collisions. Physically, the difference arises from the large mass ratio assumed to hold between the Brownian particle and the atoms, as well as the equipartition of energy among all particles and atoms in collisional equilibrium.

Here at last is the physical basis for the two time scales in the random walk in velocity space illustrated in Figures 5.11a, 5.11b, and 13.5. Because the Brownian particle is heavy, it feels rapid small impulses from the atoms, leading to a jagged velocity curve. At the same time, the atoms tend to hit more often on the forward (leading) face of the particle and this leads to a net deceleration, giving the appearance of a resistance force. The large mass ratio means that momentum transfer between the particle and the atoms is very inefficient. Only a small fraction of the particle's momentum can be transferred at each collision, hence the damping is very slow.

The fact that, by these statistical arguments, the microscopic force acting on particles in a gas can be divided into a fluctuating part and a smooth resistance is at the heart of the so-called "fluctuation–dissipation theorem." In effect, this separation provides a link between the individual collisions (which contain no appearance of friction) and the existence of friction (viscosity) in a gas. The link is a statistical one, and it depends, as we have seen, on averaging over the properties of many collisions.

13.3 PRESSURE IN AN IDEAL GAS

(a) Stationary Walls

Define the pressure p as the rate (per unit time and per unit area) at which the particles of the gas carry momentum to the walls. Equivalently, it may be considered as the rate at which the walls alter the momentum of the particles. (Note that there is no change of energy during an elastic collision with a stationary wall.) Imagine the gas to be represented by a set of beads sliding with speed v on wires strung perpendicular to the end walls, as in Figure 13.8.

We may compute the pressure as follows. The change of momentum each

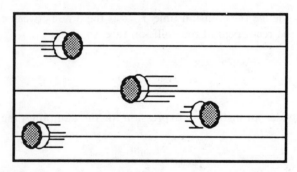

FIGURE 13.8 One-dimensional gas represented by beads sliding along wires with speed v perpendicular to the end walls, for computing pressures.

time a particle strikes the wall and rebounds elastically is $-2mv$. We add up the individual momenta brought to a unit area of the wall in a suitable time interval δt and then divide the total by τ. In the simplest case, suppose there are $C\,\delta t$ similar collisions in a time interval τ. Then the sum is given by the product

$$\sum mv = -C\,\delta t\,mv,$$

and the pressure is

$$p = 2mvC.$$

(The same average pressure ought to be obtained from both ends of the system, as long as we do not include gravitational acceleration.) To evaluate the number of collisions, $C\,\delta t$, we compute the average time interval between collisions, τ. (This implies $C = 1/\tau$.) If the wire has a length L and a total of N particles, on the average the particles are separated by a distance L/N (if $N \ll 1$), and this is the mean distance from the last particle to the end wall. The particle must traverse twice this distance between collisions, requiring a time $\tau = 2L/Nv$. Hence the number of collisions with the wall in a time δt is

$$C\,\delta t = \frac{\delta t}{\tau} = \frac{\delta t\,Nv}{2L}. \tag{13.16}$$

In place of N/L we use the average particle density, $n \equiv N/L$, and we have

$$C = nv/2, \tag{13.17}$$

$$p = mnv^2 = 2U, \tag{13.18}$$

where

$$U \equiv n(mv^2/2) \tag{13.19}$$

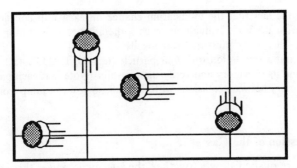

FIGURE 13.9 Representing a two-dimensional gas by beads sliding slong straight wires with speed v, for computing the relationship between pressure and density of kinetic energy. For the purpose of this calculation, the beads are assumed not to collide with each other.

is the density of kinetic energy on the wire. When atoms are not confined to motion in one dimension, the factor of proportionality between p and U becomes smaller. Suppose, for example, a two-dimensional gas is represented by beads sliding on wires strung in two directions (Figure 13.9). The pressure on each end wall remains the same because each wire only connects a single pair of end walls, but the total density of kinetic energy is doubled. In this case, $p = U$.

Exercise 13.1
Imagine a rarified gas in a two-dimensional square box to be modeled by particles with random initial positions in the xy-plane. All particles have the same speed s but random directions, and they bounce elastically from the walls but do not collide with each other (see Figure 13.10).

(a) If the box is L units on a side, show that the shortest trajectory is $2L$ (for $\alpha = 0$), while the longest is $2L\sqrt{2}$ (for $\alpha = 45°$).

(b) Evaluate the duration of a single trajectory for $\alpha = 0°$, $15°$, $30°$, and $45°$, and compute the angle α from the perpendicular to the wall at each wall collision.

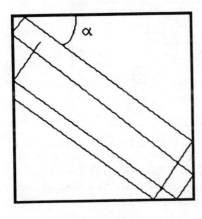

FIGURE 13.10 Two-dimensional trajectory of particle in a rectangular box making angle α with the top (the Exercise 13.1).

(c) Given the fact that the momentum change on each wall collision is $2ms \cos \alpha$ perpendicular to the wall, evaluate for each α the average rate of momentum transfer by each wall—hence the pressure on the walls.

(d) Carry out a simulation with four particles ($\alpha = 0°$, $15°$, $30°$, $45°$), following each particle around its trajectory and verify these results. (Note that each particle follows a straight trajectory between collisions with the wall, so you can jump from one collision to the next.)

(b) Compression of the Gas

If the walls are pushed slowly together, the energy increases through work done by the moving walls. Each time a particle strikes one of the moving barriers, it rebounds elastically, and the computed change of velocity must account for the motion of the wall. We accomplish this as follows.

Suppose the wall is moving to the right with velocity w. (See Figure 13.11.) A particle moving to the right and striking the wall with speed v rebounds with velocity $-v + 2w$. (This result may be derived by considering the collision in a coordinate system in which the wall is at rest and then transforming back to the original coordinate system. Each transformation adds velocity $-w$ to the particle's velocity.) The total change of the particle's velocity is therefore $-2(v-w)$ at each such collision.

If $w < 0$, the right-hand wall is moving toward the particles, and the particles' speeds are increased on collision. The density of the particles will increase as the volume decreases and so will the frequency of collisions with the wall. These

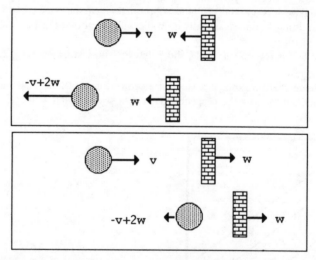

FIGURE 13.11   Ball rebounds from moving wall with altered velocity, depending on the wall's motion. In the upper frame, the wall is moving toward the ball at the time of impact and the ball rebounds swiftly. In the lower frame, the wall is moving away from the ball at the time of impact and the ball rebounds slowly. This behavior causes the gas in a contracting enclosure to become hotter.

effects will increase the pressure. We ask: What is the relationship between the volume of the gas and the pressure?

Exercise 13.2
Simulate a one-dimensional gas by a single particle flying back and forth in a cylinder of length V and bouncing elastically from the ends. Slowly shorten the tube and verify that the particle's speed increases according to the power law, $v \sim L^{-\gamma}$. (This is most easily done on a log–log plot of ln v against ln L, which yields an approximately straight line whose slope is $-\alpha$.) Evaluate the exponent γ, and compare it with the theoretical value discussed below.

If all motions are confined to a single direction, we may predict the outcome of the compression as follows. The time between collisions with the moving end is $\tau = 2L/v$ and the pressure is $p = 2mv/\tau = mv^2/L$. We want to calculate how rapidly v and the pressure p increase as the cylinder is shortened. Figure 13.12 shows a simulated gas cylinder in which a piston rises and falls, changing the volume, pressure, and temperature of the gas, which consists of a "cloud" of four atoms.

To estimate the influence of the piston's motion, suppose the piston moves inward with speed r, and the particle speed increases by an amount $2r$ on each collision. (This assumes that the piston is massive.) The average rate of increase of speed will be

$$\frac{dv}{dt} = \frac{2r}{\tau} = \frac{rv}{L}. \tag{13.20}$$

With $dL/dt = -r$, this gives

$$\frac{1}{v}\frac{dv}{dt} = -\frac{1}{L}\frac{dL}{dt}$$

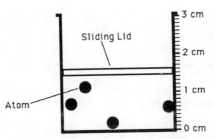

FIGURE 13.12   Simulation of a four-atom gas in a cylinder held by a sliding piston whose weight is balanced by the impacts of the moving atoms. When the piston moves downward, the atoms receive kinetic energy, and the gas becomes hot. At the same time, the pressure of the atoms increases so the motion of the piston is reversed. In a simulation like this, the piston is constantly moving up and down in a haphazard fashion. Figure 13.13 shows the relationships between pressure and volume in such a gas.

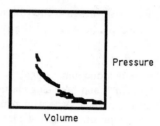

FIGURE 13.13   Pressure versus volume in a cylinder as the piston moves down. Here the pressure is defined as $\sum mv^2$/volume, where m and v are the atom's mass and speed. The volume of the cylinder inside the piston is $L\sigma$. This diagram indicates that the pressure increases as the volume decreases.

or

$$d \ln v = - d \ln L.\tag{13.21}$$

This implies that a plot of ln (speed) against ln (length) will be a straight line with a slope $\alpha = -1$. (Does this agree with the result of the simulation in Exercise 13.2?)

By taking the logarithm of (13.18) for pressure and differentiating, we find

$$\frac{d \ln p}{dt} = 2 \frac{d \ln v}{dt} - \frac{d \ln L}{dt},$$

so with (13.21)

$$\frac{d \ln p}{dt} = - 3 \frac{d \ln L}{dt}.\tag{13.22}$$

We may write the integral of (13.22) as

$$p \sim L^{-\gamma} \qquad (\gamma = 3).\tag{13.23}$$

Thus, the pressure and volume of the gas are related by a power law, and the system would trace a hyperbola on the PL-plane (Figure 13.13). Figure 13.6 qualitatively verifies this relation for one particular simulation. (The value of the exponent becomes $\gamma = \frac{5}{3}$ for a three-dimensional gas.)

(c) First Law of Thermodynamics

If the lid in Figure 13.12 slides downward and compresses the gas, we may find an important relationship between the kinetic energy of the atoms and the work done by the piston. Suppose for simplicity that the piston (whose volume is the product of its length and cross-sectional area, σL) contains a single atom of mass m and speed v. The lid is moving downward at speed dL/dt. The kinetic energy of the particle is $mv^2/2$, and this increases as the atom bounces off the moving lid. The rate of change of the kinetic energy of a single particle is

$dKE/dt = mv(dv/dt)$. But by considering the collisions in detail, we have just found $dv/dt = -(v/L)dL/dt$, and combining these expressions, we have for a single particle

$$\left(\frac{dKE}{dt}\right)_{particle} = -\left(\frac{mv^2}{L}\right)\frac{dL}{dt}, \qquad (13.24)$$

and for the particles inside the piston

$$\frac{dKE}{dt} = -N\left(\frac{mv^2}{L}\right)\frac{dL}{dt}. \qquad (13.25)$$

For the pressure we have $p = mNv^2/(\sigma L)$, so

$$\frac{dKE}{dt} = -p\sigma\frac{dL}{dt}. \qquad (13.26)$$

The right-hand side of (13.26) is just the rate at which the lid does work on the atom, and it gives the first law of thermodynamics in a restricted form: *If there is no other form of energy exchange through the walls of the container, the change in thermal energy of a gas equals the work done by the walls.*

(d) Stability of the Gas

According to this model, the pressure beneath the lid will increase as the lid comes down, because v^2 increases and the volume decreases. Thus compression of the gas requires more and more force as the volume gets smaller. If the downward force is constant, there will come a point where it is overcome by the upward force. At this moment the lid will start accelerating upward. Ultimately, the gas will expand, and the pressure will start to drop. This situation is *stable*, in the sense that the gas will not collapse on itself.

If, by contrast, the downward force on the lid were to increase more rapidly than the pressure, the lid would continue to accelerate downward. The system would then be unstable and would collapse. This can occur in the deep interiors of stars, where gravity pulls the gas downward. This type of instability is thought to be the cause of nova explosions.

REFERENCES

Atkins 1984
Baker 1986
Boyd 1989
Dempsey and Hartman 1986

Gould and Tobochnik 1988
Hill 1963
Koonin 1986
Rechtman 1988
Reichl 1980
Rio et al. 1976
Sherwood and Bernard 1984
van Ness 1983

14

UNIFORM GAS OF ATOMS AND PHOTONS

In this chapter, we endow atoms with internal energy and permit them to emit and absorb photons. This leads us to study the equilibrium of photons and atoms in an enclosure. With the goal of running these simulations on microcomputers, we make several substantial simplifications that limit us to uniform gases.

14.1 COLLISIONS AMONG EXCITED ATOMS

We assume atoms can be excited to a discrete set of internal energy levels by acquiring energy in collisions with other atoms. Conversely, an atom may descend to lower levels during a collision by the transformation of internal energy to kinetic energy. We ask for the average distribution of atoms among its energy states.

Because we deal with a uniform gas, we do not need to follow the atoms' trajectories through space. Atoms are identified by an index, and their only attributes are kinetic energy and excitation energy. We simulate atom–atom collisions by randomly selecting pairs of atoms and causing them to share energy.

The energy of an atom will consist of two parts: kinetic energy K and excitation energy I. The kinetic energy can take any positive value, while the excitation energy levels are discrete and will be assumed to follow a simple expression of the type

$$I_n = \chi \left(1 - \frac{1}{(n+1)^a} \right) \qquad (n = 0, 1, 2, \ldots).$$

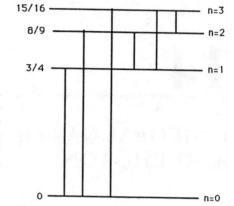

FIGURE 14.1 Energy levels I_n of a simplified atom, labeled according to excitation energy measured from the lowest level. The vertical lines indicate transitions among levels, corresponding to changes of excitation energy. In this case ($a = 2$) the energies are given by $I_n = 1 - 1/(n + 1)^2$.

With $a = 2$, we have $I_n = 0$, $3\chi/4$, $8\chi/9$, $15\chi/16, \ldots$, as for hydrogen. The energy-level diagram for a four-level atom of this type is shown in Figure 14.1, with the six possible transitions marked. (The number of transitions is the number of combinations of four excitation levels taken two at a time.)

We assume the collection of atoms to be isolated in an enclosure of field-free space, and we seek the average (equilibrium) description. Each atom is tagged with an index number $(1, \ldots, N)$, and the data for the atom consist of its kinetic energy and the level to which it is currently excited. The simulation will proceed as a repeated sequence of the following steps:

Step 1: Randomly select two integers from the range 1 to N, designating the atoms that are to collide. Call them atom-1 and atom-2.

Step 2: Draw (either randomly or cyclicly) an integer, p, from the range 1–3 to select the energy-exchange process for this collision from the following list (see Figure 14.2):

$p = 1$ *Elastic collision*:   randomly redistributing the atoms' kinetic energies between the pair;

$p = 2$ *Excitation* of atom-1, subtracting kinetic energy from atom-2;

$p = 3$ *De-excitation* of atom-1, adding kinetic energy to atom-2.

Step 3: If $p = 1$, select a random fraction of the kinetic energy of atom-1 and give it to atom-2. This is an elastic collision and it does not affect the excitation of the atoms.

If $p = 2$, we deal with an excitation in which some of the kinetic energy of atom-2 may be used to raise atom-1 to a higher level of excitation. We first verify that atom-2 has sufficient energy to induce a transition in atom-1, or else we return to step 1. This is accomplished by computing which levels can be reached from the current level of atom-1. (We exclude excitations upward from the top level, as we do not wish to deal with

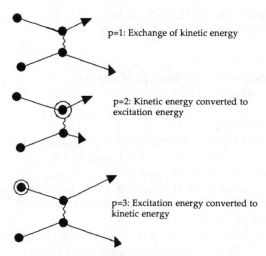

p=1: Exchange of kinetic energy

p=2: Kinetic energy converted to excitation energy

p=3: Excitation energy converted to kinetic energy

FIGURE 14.2 Schematic representation of three types of collision between a pair of ground-state atoms (indicated by solid dots) or an atom and an excited atom (indicated by circled dots). $p = 1$: energy is exchanged only as kinetic energy; $p = 2$: some kinetic energy is converted to excitation energy; $p = 3$: some excitation energy is converted to kinetic energy. The simulation considers all three types of collision.

ionization at this stage.) One of these levels is selected randomly (or according to a table of collisional transition probabilities) and the exchange is carried out by updating the data on each atom. This consists of subtracting the required energy from the kinetic energy of atom-2 and altering the excitation level of atom-1. We also record the two energy levels of atom-1 that were involved in this process.

If $p = 3$, some of the excitation energy of atom-1 is given to atom-2. This is accomplished by selecting a lower level of atom-1 randomly (or according to a table of spontaneous transition probabilities), computing the energy difference and adding this to the kinetic energy of atom-2. We then update the data on the atoms and record the two levels in atom-1.

The rules governing the selection of excitation levels in step 3 will determine the relative frequencies of various types of processes. In an equilibrium system, the transition probabilities do not influence the average distribution functions, as we shall see.

The following program fragments show the handling of the three types of energy exchange discussed above: elastic collision, collisional excitation, and collisional de-excitation.

Procedure ElasticColl (a1, a2 : integer;

    var   deltE : real); {gives random fraction of atom1 kin energy to atom2}
{returns deltE}
    Begin
    deltE := abs(random / 32767) * atom[A1].KinE;

```
        atom[a1].kinE := atom[a1].kinE - deltE;
        atom[a2].kinE := atom[a2].kinE + deltE;
        End;


Procedure Excitation (a1, a2 : integer; tl : integer;
        var L1, L2 : integer;
        var e : real;
        var jump : boolean);
{call with indices of selected atoms and highest excitation level}
{returns initial and final levels of atom1, energy change}
{and jump := true if a jump took place }
{atom1 receives energy from atom2; find levels acessible to atom1}
{given energy of atom2, then select one of them randomly}
{ionization is not allowed}
{returns initial and final level designators for atom 1 and deltE}
{or jump := false, indicating no move was possible}
        Begin
        jump := false;
        e := 0;
        If Not (atom[a1].level = tl) Then   {do not jump upward from top level}
                Begin
{find what levels are available}
{store level and find how high atom2 can raise atom1}
                L1 := atom[a1].level;
                L2 := L1;  {initialize}
                Repeat
                        L2 := L2 + 1
                Until (L2 = tl)
                        Or ((levelEnergy[L2] - LevelEnergy[L1]) > atom[a2].kinE);
{randomly select upper level}
                If(levelEnergy[L2] - LevelEnergy[L1]) > atom[a2].kinE Then
                        L2 := L2 - 1;
                L2 := randInt(L1, L2);
                If L2 > L1 Then
                        Begin
                        jump := true;  {find change of energy, update atom}
                e := levelEnergy[L2] - levelEnergy[L1];
                        atom[a1].level := atom[a1].level + (L2 - L1);
                        atom[a2].kinE := atom[a2].kinE - e;
                        End;  {if }
                End;  {if not}
        End;  {Excitation}


Procedure DeExcitation (a1, a2, tl : integer;
        var L1, L2 : integer;
```

```
      var  e : real;
      var  jump : boolean);
{atom1 decends and gives excitation energy to atom2 as kinetic energy}
      Begin
      jump := false;
      e := 0;
      If atom[a1].level > 0 Then  {jump only if in excited level}
          Begin
          jump := true;
          L1 := atom[a1].level;  {initial level}
          L2 := randint(0, L1);  {randomly select lower level from 0 to L1}
          e := levelenergy[L1] - levelenergy[L2];  {find energy change}
          atom[a1].level := L2;  {put atom1 in final state}
          atom[a2].kinE := atom[a2].kinE + e;
{update kinetic energy of atom2}
          End;
      End;  {DeExcitation}
```

The results of a sample simulation using this algorithm are shown in Figures 14.3a and 14.3b.

14.2   GENERALIZED DESCRIPTION OF A UNIFORM GAS

Consider an isolated collection of N infinitesimal particles of mass m, which share a prescribed amount of energy. Suppose that this energy may appear in several forms, that is, kinetic energy of motion, K, potential energy of position, V, and internal energy of excitation, I. The kinetic energy of a particle of speed v is, according to classical mechanics, $K = mv^2/2$; the potential energy at height h in a uniform gravitational field is mgh. We shall assume the atoms to move according to Newton's laws. The excitation energy is independent of position or speed and depends only on the atom's internal structure. It can only take discrete values, and each value is associated with a "level of excitation," n. The energy of the nth level is written I_n.

The atoms collide with each other and randomly exchange energy as well as altering the energy from one form to another. For example, two atoms may collide and increase the excitation of one or both while decreasing the sum of their kinetic energies. In such interactions the total energy is conserved, so the sums over all atoms before and after the collision are the same:

$$\sum I_n + \sum K = \sum I'_n + \sum K'. \tag{14.1}$$

We need not include the potential energy because this is a function of position, which is assumed not to change significantly during a collision.

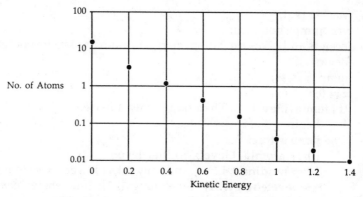

FIGURE 14.3a Average distribution of kinetic energies of colliding atoms for a simulation as described in this section, computed with the following parameters: $N_{atoms} = 20$, total energy of system $= 0.4$ units. To construct this histogram, atoms were binned into intervals of 0.2 in kinetic energy, atom[i].kinE. On this semilogarithmic plot the straight-line decrease in the number of atoms with increasing energy indicates an exponential (Boltzmann) distribution. The slope of the relationship is a measure of $1/T$, the reciprocal of the temperature of the gas. For a colder gas, the slope would be steeper.

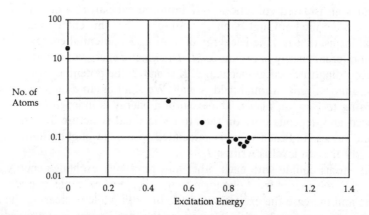

FIGURE 14.3b Distribution of excitation energy among colliding atoms for the simulation of Figure 14.3a. As in the previous figure, the straight-line decrease of the number of atoms in levels of increasing energy indicates an exponential distribution. The slopes of the relationships in the two figures are approximately the same. The deviation from a purely exponential relation shown among the highest three energy levels is caused by the finite number of levels in the simulated atom.

Now let us define three time-averaged distribution functions, one for each type of energy:

$P_K dK$ = probability that an atom's kinetic energy lies in the interval $(K, K + dK)$;

$P_V dV$ = probability that an atom's potential energy lies in the interval $(V, V + dV)$;

P_n = probability that an atom is in the nth level of excitation.

The distribution functions for K and V are continuous, while that for excitation is discrete. The probabilities obey the following normalizations:

$$\int P_K dK = \int P_V dV = \sum_n P_n = 1 \qquad (14.2)$$

and the number of particles in each energy range is related to the total number N by

$$N_K dK = N P_K dK, \qquad (14.3a)$$

$$N_V dV = N P_V dV, \qquad (14.3b)$$

$$N_n = N P_n. \qquad (14.3c)$$

We now define a set of rates to describe the processes by which atoms exchange energy. The processes are of two types:

(1) Collisions in which an atom alters its kinetic energy. We define $R_{KK'}$ as the rate at which atoms with energy in the range $(K, K + dK)$ jump to the range $(K', K' + dK')$. We write this as proportional to the number of atoms in the initial state, so

$$R_{KK'} = N_K C_{KK'}. \qquad (14.4a)$$

The rate of the inverse is written

$$R_{K'K} = N_{K'} C_{K'K}. \qquad (14.4b)$$

The coefficients $C_{K'K}$ and $C_{KK'}$ depend on the rate at which the atoms collide with each other and the efficiency of the collisions to transfer energy. Ordinarily, they are computed from detailed quantum mechanical theory, but we substitute simple models.

(2) Collisions in which an atom alters its excitation energy. Similarly, we define $R_{nn'} = N_n C_{nn'}$ as the rate at which an atom in level n jumps to level n'. The inverse rate is $R_{n'n} = N_{n'} C_{n'n}$.

We will confine our attention for the moment to systems that are isolated and at constant energy. We ask for the average behavior of the atoms. Three important rules governing such a system can be described. They are the following:

(1) *Distribution Function*: The shapes of the average distribution functions are independent of the manner of shuffling, as long as the shuffling is complete and does not exclude any energy states. The distribution functions obey the Boltzmann formula,

$$P_K \sim g_K \exp\left(-\frac{K}{kT}\right), \qquad P_V \sim \exp\left(-\frac{V}{kT}\right), \qquad P_n \sim g_n \exp\left(-\frac{I_n}{kT}\right). \quad (14.5)$$

(2) *Equipartition of Energy*: In a gas of colliding atoms, the mean kinetic energy of each particle is the same, regardless of the particle's mass. Heavy particles therefore tend to move more slowly.

(3) *Detailed Balancing*: The rate of each energy-exchange process is equal to the rate of the corresponding inverse process:

$$R_{KK'} = R_{K'K}, \quad \text{exchanges of kinetic energy}, \qquad (14.6a)$$

$$R_{nn'} = R_{n'n}, \quad \text{exchanges of internal energy}. \qquad (14.6b)$$

This is the principle of detailed balancing for an equilibrium system, and the array in Table 14.1 shows an example drawn from the simulation illustrated in Figure 14.3. In this array, we tabulate the number of the transitions among energy levels in the excited atoms, $R_{nn'}$. The off-diagonal terms at similar positions ought to balance, but they do not quite do so, as we did not carry out the simulation long enough to overcome the effect of small numbers. The right-hand column and the bottom row show the sums, and they represent the

TABLE 14.1   Number of Transitions among Excited States, $R_{nn'}$

n	$n' = 0$	1	2	3	4	5	6	7	8	9	Sum
0	0	152	39	17	8	9	6	9	13	1	254
1	171	155	28	12	3	8	10	2	1	2	392
2	33	39	31	8	2	8	1	3	0	3	128
3	19	8	14	21	9	9	5	2	2	1	90
4	7	8	5	10	9	6	4	4	2	3	58
5	8	10	4	6	6	8	8	7	1	2	60
6	3	6	3	9	5	6	4	8	5	4	53
7	5	6	2	3	0	2	8	3	9	4	42
8	5	3	1	2	8	3	4	0	4	11	42
9	1	6	1	2	8	1	3	4	5	7	38
Sum	252	393	128	90	58	60	53	42	42	38	

total number of exits and entries to the corresponding levels:

$$\sum_{n'} R_{n'n} \quad \text{and} \quad \sum_{n'} R_{nn'}. \tag{14.7}$$

In an equilibrium enclosure these numbers ought to be equal, and they are nearly so in this case.

These three rules work well for gases that are like Goldilocks' porridge: not too hot, not too cool, not too thick, not too thin. In more technical terms, these models are appropriate for warm, "dilute" gases in which the atoms (1) are very small compared to the average distance between pairs of atoms (so they do not squeeze each other), (2) are moving much slower than the speed of light (so relativistic corrections are unimportant), and (3) are warm enough so that the quantum zero-point energy is not important.

14.3 DETAILED BALANCING IN EQUILIBRIUM

The principle of detailed balancing in equilibrium (Chapter 13) is an extremely useful and important concept. To prove it rigorously from first principles is a difficult and not particularly interesting task, so we shall adopt it as an axiom and demonstrate its validity.

If we turn question around, it is easy to see that detailed balancing is a *sufficient* condition for the distribution functions to become independent of time. This follows from the fact that all the net rates vanish, so all the populations must be constant. For example, the rate at which the population of the nth level of excitation changes is given by the difference between inflow and outflow.

$$\frac{dN_n}{dt} = \sum_{n'} (R_{n'n} - R_{nn'}), \tag{14.8}$$

where $R_{nn'}$ is the total rate off transitions (collisionally and radiatively induced) from n to n'. But if detailed balancing holds, every pair of terms on the right-hand side vanishes, so the summation vanishes and the populations are independent of time.

14.4 MIXTURE OF ATOMS AND PHOTONS

In this section we simulate a more complex system, including photons as well as atoms. We return to the approach developed in Section 14.1 and assign an initial kinetic energy and excitation state to each atom. We then follow the energy and excitation history of each atom.

We will again deal with a homogeneous medium, which has translational invariance, so we do not need to follow the individual trajectories of the particles.

Elastic and inelastic collisions between atoms will be treated as in Section 14.1, and we extend the formulation by adding spontaneous emission of photons (with a concomitant de-excitation of an atom), photoelectric excitations by photon absorption, and stimulated emissions (which act as negative absorptions).

The atom model has four excited energy levels computed with $a = 1$, so $I_n = 1000[1 - 1(n + 1)]$, with $n = 0, 1, \ldots, 4$. The factor 1000 is arbitrary and was chosen to permit representing the energies by convenient integers to speed the calculation. The energy levels are 0, 500, 667, 750, 800, 833, and differences between these values correspond to possible photon energies.

Each photon is characterized by a single parameter, PEnergy. This energy is given by the difference in energies of the upper and lower atomic excitation levels involved in the transition, $PEnergy = I_U - I_L$. (The fact that the photon energies in the spectrum of an atom can be derived from the differences of a smaller number of terms, I_n, is known as the "Ritz combination rule." It was discovered empirically before the Bohr model of quantized levels was developed.)

Three additional processes, corresponding to $p = 4$, 5, and 6, are added to the list, as follows:

$p = 4$: *Spontaneous Emission:* An atom is selected randomly by assigning an integer value to **a1**. If the atom is in an excited state, it makes a transition to a randomly chosen lower state and produces a photon whose energy is the energy difference. The boolean expression **jumps** is set to **true**. (In a more sophisticated simulation, the transition occurs according to probabilities of transition to various levels.) After the value of **a1** has been selected, the following program fragment performs this calculation.

```
Procedure SpontEm;
        var  i : integer;
        Begin
        jump := false; {initialize to null transition}
        L1 := atom[a1].level; {find initial level}
        If L1 > 0 Then   {atom is excited, spontaneous emission is possible}
                Begin
                L2 := randint(0, L1 - 1);  {pick lower level}
{here is place for probabilities }
                jump := true; {indicate that a transition ought to be recorded}
                atom[a1].level := L2; {place atom in new lower level}
                totPhot := totPhot + 1; {increment total photon count}
                With histogram[photonE] Do  {place new photon in bin}
                        Begin {Compute energy and find bin for new photon}
                        i := (levelEnergy[L1] - levelEnergy[L2]) Div interval;
{interval is the bin size; increment bin count}
                        count[i] := count[i] + 1;
                        End {with}
                End; {if}
        End; {SpontEm}
```

The number of photons **totPhot** is a variable so a dynamic storage would be appropriate, rather than simply using an array of specified size. The photons might be stored as a linked list, and in this case each individual could be traced through time. A simpler method is merely to bin the photons (as is done in the program fragment above) and keep track of the numbers of atoms of different energies. Doing this, we lose track of individual photons. This is acceptable in many calculations for a homogeneous medium, but as we shall soon see, it adds a slight complication to the treatment of absorptions and stimulated emissions.

$p = 5$: *Photoabsorption:* An atom is selected randomly from the list of atoms. If the atom is not in its highest permissible level (we ignore the possibility of ionization), a photon is selected randomly from the bins in a two-step process. First, a number, **PNum**, is selected randomly from the interval 1 to **totPhot**. This number identifies the photon in a ranked list of photon energies. It may be selected by computing **PNum = trunc(fract∗totPhot)**, where **fract** is a random number in the range 0–1 and **trunc** indicates the integer part. Second, the bin counts are added sequentially until the cumulative count **cumCount** exceeds **PNum**. A photon is selected from the corresponding bin and assigned an energy equal to the mean energy of the bin. These two steps are achieved by the following procedure.

```
Procedure PickPhoton;
        var        fract : real;
                   PNum, bin, cumcount : integer;
        Begin
        fract := abs(random / 32767); {random fraction to pick photon}
        pNum := trunc(totPhot * fract);
{compute corresponding serial number of photon}
        bin := -1;
        cumcount := 0;
        With histogram[PhotonE] Do
                Begin
                Repeat
{cumulate bin counts until pNum is achieved or last bin counted}
                bin := bin + 1;
                cumCount := cumCount + count[bin]
                Until (cumcount > pNum) Or (bin > 49);
                pEnergy := interval * bin + interval DIV 2;
{assign mean bin energy}
                End; {with}
        End; {pickPhoton}
```

Next, the procedure **CheckJump**, determines whether the photon energy PEnergy corresponds to excitation to an upper level. That is, it sees whether $I_U - I_L = $ PEnergy for any U, given that the atom is currently in state L. If so, it returns the boolean expression **Jumps:= true** and the procedure **PEAbsorb** then updates the photon statistics and places the atom in its new level. The bin count is decreased by unity, as is **totPhot**, the total number of photons. These steps are indicated in the following procedures.

Procedure checkJump (a1, PEnergy, TL, L1 : integer;
         var   **L2 : integer;**
         var   **jump : boolean);**
{Jump := true if photon energy corresponds to a jump from Level 1}
         var
                lineWidth, i : integer;
         Begin
         lineWidth := histogram[photonE].interval DIV 2;
{PEnergy must be within a line width of level difference}
         jump := false; {initialize}
         i := L1; {Start search from current level and work upward}
         While (i < tl) And Not jump Do
           Begin
{look for level difference that is within HalfInterval of PEnergy }
{quit when condition is true or scanned all upper levels}
{ only one upper level is available with given PEnergy, search can stop}
           i := i + 1;
           jump := ((PEnergy + lineWidth) >=
           (levelEnergy[i] - levelenergy[L1])) AND ((PEnergy - lineWidth)
           <= (levelEnergy[i] - levelenergy[L1]));
{Jump is true when condition is met.}
         End; {while}
         IF jump Then L2 := i; {Index of new level. }
         End;

Procedure PEAbsorb; {Carry out the process and update statistics}
         var   **bin : integer;**
         Begin
         jump := false;
         L1 := atom[a1].level;
         If L1 < tl Then {can be absorbed only by atom below top level}
           checkJump(a1, pEnergy, tl, L1, L2, jump);
         If jump Then {update the statistics}
           With histogram[photonE] Do
           Begin
           bin := pEnergy Div interval; {find bin number of photon}
           count[bin] := count[bin] - 1; {remove photon from the bin}
           totPhot := totPhot - 1; {remove a photon from distribution}
           atom[a1].level := L2; {update atom record}
           End;
         End; {PEAbsorb}

$p = 6$: *Stimulated Emission:* This is similar to **PEAbsorb** and is treated as a negative absorption. That is, after an atom and a photon have been selected and it is ascertained that the atom is not in the ground state, **checkJump** looks for a lower level whose energy difference equals **PEnergy**. If it finds one, the program creates a new photon of the same energy and updates the statistics.

The results of one such simulation for a five-level atom, are shown in Figures 14.4 and 14.5. Figure 14.4 shows the atom kinetic energy distribution as a histogram, with the familiar drop off toward higher energies. Initially, the gas consisted of 40 atoms in the ground state with kinetic energies = 2000, in arbitrary units. There were no photons initially, so all photons were created by spontaneous and stimulated emissions, and they each had the energy of a level transition.

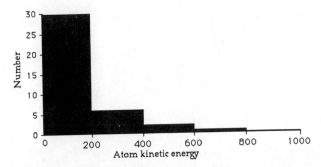

FIGURE 14.4   Histogram of atoms' kinetic energies after a simulation of about 60,000 transitions in a gas consisting of atoms and photons. The shape is characteristic of the Boltzmann distribution expected for an equilibrium gas.

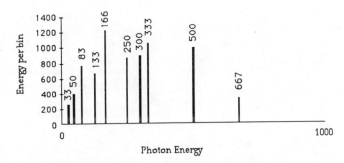

FIGURE 14.5   Spectrum of photons, in which the energy per bin is plotted against the bin energy. The level energies for this atom were 0, 500, 667, 750, 800, and 833. Each photon energy (indicated above the histogram) corresponds to the difference between at least one pair of level energies. Spectrum lines at PEnergy = 133 and 167 correspond to two distinct transitions, as the reader should verify.

Such a program might also keep track of each type of transition, in the form of arrays C_{if} giving the number of transitions from state i to f induced by collisions and R_{if} giving the number induced by photons. The symmetry (or lack of symmetry) of this array will indicate the extent to which detailed balance was achieved and is a check on the validity of the simulation.

REFERENCES

Crawford 1988
Vincenti and Kruger 1965

15

MODELING PLANETS AND STARS

Stars and planets form out of condensations from gas and dust scattered in space. Stars are self-heating and gaseous; they later return most of their matter to space, where it may again condense and form the next generation of stars. Planets are solid or liquid and they generate very little heat. A planet's surface is usually covered by a thin layer of gas that forms its atmosphere.

This chapter describes simple simulations of nonuniform gases such as planetary atmospheres and the interiors of stars. The nonuniformity has three principle causes:

(1) The force of gravity tends to separate atoms into layers of different density and to make the heavy atoms sink lower than the light atoms. Diffusion (random walk) of atoms tends to overcome this separation.

(2) The production of heat in the interior, either by nuclear burning or by compression of the gas, tends to make the central regions of a star hotter and more adept at supporting the external layers.

(3) The stellar surface, from which photons may escape and carry away the heat of the star, tends to cool the outer layers.

We will investigate the first two causes in this chapter.

15.1 IRREVERSIBILITY AND THE MIXING OF ATOMS BY DIFFUSION

If two types of atoms are enclosed in a box and there is no gravitational field, each atom will fly randomly about, colliding with other atoms and gradually

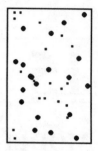

FIGURE 15.1 Light and heavy atoms are initially segregated into two halves of a box. After a short interval of random walking, they have diffused through the box and each type will have a uniform likelihood of being found anywhere in the box.

Starting positions   After diffusion

moving away from its starting point. There is no preferred direction of motion, and, as a result, the two types will tend to spread uniformly through the box. This is the process of diffusion, and it accounts for the spread of aromas in a room. Figure 15.1 shows the result of a diffusion simulation using the program of Chapter 13 for a gas in a box. The left side shows the initial positions: heavy atoms to the right and light atoms to the left. The right side shows the situation after the atoms have had sufficient time to "forget" their starting points by wandering through the box.

This is an example of an *irreversible* process. It gets it name from the fact that it will not work backward if there are enough atoms in the box. That is, if we start with the atoms uniformly scattered, there is an extremely small chance that they will segregate into the two sides of the box. (If there were only 2 atoms of each type, they would be separated, heavy on right and light on left, about 1/4 of the time.)

Exercise 15.1

(a) Show that the probability of such segregation into two halves is $(1/2)^N$ if there are N particles.

(b) Carry out a simulation with two types of particle and verify this expectation by keeping track of the times when they are segregated.

To express the condition for irreversibility another way, suppose you were shown the two snapshots in Figure 15.1. You would know at once that the one on the right was made *after* the one on the left. That is, you can infer the direction of the passage of time (*time's arrow*) simply by inspecting the two figures. On the other hand, if you were shown two snapshots of a box containing only 2 atoms, you could not be sure which came first. Identifying the arrow of time and irreversibility requires having a large number of atoms in the system.

15.2   DENSITY LAYERING IN AN ATMOSPHERE

(a) Pure Gas

We start with a flat planetary atmosphere in a gravitational field. The gas is comprised of a single type of atom, so all particles have the same mass. We imagine the atmosphere to be confined in a motionless box with perfectly elastic, opaque walls. When an atom strikes the wall, its velocity perpendicular to the wall is reversed in direction, but its magnitude is not changed.

We may simulate this system with the same program that was used for the uniform gas in Chapter 13, with a single alteration. We put the box into a constant gravitational field so each particle is accelerated downward by the same amount. During a time-step δt, each atom's vertical speed is changed by an amount $-g\,\delta t$, while the horizontal speed is not affected. Thus the atom acts like a projectile in a parabolic orbit between collisions. Each collision is assumed to be an instantaneous event in which the kinetic energy and momentum are shuffled between the numbers of the colliding pair while the total energy and momentum are conserved. Gravity has no effect during the brief moment of a collision. We assume that collisions only take place between two atoms at a time.

Because the atom's vertical speed is not constant between collisions, we must be a little careful computing how far it moves during one time-step. Computing the horizontal displacement is straightforward: we merely take $x \rightarrow x + v_h\delta t$ because v_h is constant. We may use a similar expression for the vertical displacement, $y \rightarrow y + v_v\delta t$, if we make the time-step sufficiently small so the vertical velocity v_v can be considered constant during the interval.

The results of such a simulation are shown in Figure 15.2. The positions at a particular instant are shown on the right, while the time-averaged histograms of numbers of atoms with different speeds and heights are shown on the left.

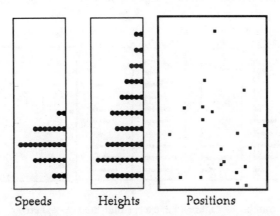

Speeds   Heights   Positions

FIGURE 15.2 Simulation of pure gas in a constant gravitational field confined to a box with elastic walls. A snapshot of the spatial positions of the atoms is shown on the right; histograms of the time-averaged speed and height distributions are also shown. As described in the text, the atoms all have the same mass and they collide with each other and the walls elastically (i.e., with conservation of energy and momentum). They are accelerated downward by a constant gravitational field. This imitates a small portion of a planetary atmosphere consisting of a single type of gas.

These averaged histograms were computed by adding together a series of histograms computed every 20 time-steps. (There is no need to compute them more frequently as they change relatively slowly.)

Exercise 15.2
Prepare a computer program to carry out the simulation described in this section in such a way that the atom–atom collisions may be suppressed. Then suppose you were to divide the atmosphere into horizontal layers and compute the mean speeds in each layer.

(a) How do you predict the mean speeds will vary with height?

(b) Perform the calculation and show that these speed are the same at all heights regardless of whether or not collisions take place. This odd behavior illustrates the tendency of an atmosphere to have a constant temperature if it is confined in a box.

(b) Mixed Gas with Two Types of Atom

Now we consider a gas consisting of two types of atoms, identical except for their masses. As before, we let them collide elastically with each other and with the walls of the box. Gravity accelerates them all downward by the same amount. How should we expect the two types of atom to behave?

We can crudely estimate the behavior by recalling the discussion of the Brownian particle in Chapter 13. In a collision between a heavy atom and a light atom, the kinetic energy will tend to be divided equally, so the heavy atom will, on average, rebound less rapidly than the light atom. Thus the light atoms will tend to acquire a greater average speed. Furthermore, the height to which an atom can fly against gravity will depend only on the square of its speed, and not at all on its mass (height $\sim v^2/2$). The results are shown in Figure 15.3,

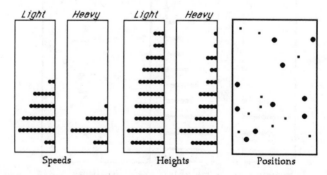

FIGURE 15.3 Similar to Figure 15.2, except that the atmosphere is composed of two types of atom, and the behavior of each type is shown in a separate histogram. The heavier atoms are assumed to be four times as massive as the lighter atoms. All atoms can collide elastically with each other and with the walls. Note that the speeds of the lighter atoms are about twice as great as the speeds of the heavier atoms (corresponding to equipartition of energy). This higher speed permits the lighter atoms to move higher in the atmosphere, so their height histogram falls off more slowly with increasing height.

where we see the greater speed and the greater average height of the lighter atoms.

Exercise 15.3
Show, by a simulation similar to the one in Exercise 15.2, that the heavy and light atoms attain the same mean energies after they have adequate time to collide with each other.

15.3 THE SPRINGINESS OF STARS

All stars are intrinsically variable in one way or another, although the variations of most stars are too subtle to be detected from Earth. The most spectacular type of variation—a supernova explosion—probably occurs only once in a stellar lifetime. Others, the cataclysmic variables, explode cyclicly, and the star seems little changed after each explosion. Another group swings gently up and down in brightness with a precise period. The remainder of this chapter will focus on this, the simplest type of variable star—known as a pulsating variable.

The brightness changes of pulsating variables are caused by the star's growing and shrinking periodically while remaining nearly spherical. These pulsations represent outward and inward swings of shells of gas, which behave like weights on a spring. In order to model this motion, we consider the atoms of the star to be grouped in several regions separated by elastic shells that act as impenetrable layers. Each shell contains a small number of atoms, and the history of each atom is followed in detail. Despite the simplicity of these simulations, their behavior is physically realistic and gives quite an accurate replication of the behavior of an actual star—far better, in fact, than might be anticipated. By confining ourselves to one or two dozen atoms, we can carry out these computations on a microcomputer.

Heat energy is largely trapped in the interior, and we may start by ignoring the effect of photons. In its simplest form, the simulation assumes the atoms move radially inward and outward until they strike the shell or a small elastic sphere at the center of the star. By thus confining the atoms to move directly inward and outward along the radius of the star, we speed up the calculations with little loss of physical reality.

The first question is: What happens to the speeds of the atoms colliding with the shell as a shell contracts? We have already simulated this aspect when we considered the compression of a uniform gas in Chapter 13. We can say that, if the shell is moving toward the atom, the atom will speed up. This means that the behavior of atoms striking from the inner face will depend on the motion of the shell, as illustrated in Figure 15.4.

When a shell contracts, the pressure on its inner face increases, and the inner atoms push back more vigorously. The pressure on the outer face decreases, and the outer atoms push less vigorously. So the interior of the star tends to resist the contraction. By a similar argument, the outer shell will resist expansion

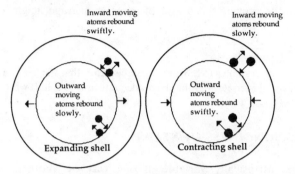

FIGURE 15.4   The star model consists of atoms divided by shells and this illustrates the effect of shell motion on the speeds of the atoms. If the shell is expanding (left), atoms striking from the inside will rebound slowly, while those striking from the outside rebound swiftly. If the shell is contracting (right), atoms striking from the inside will rebound swiftly, while those striking from the outside rebound slowly.

of the interior. The acceleration of each shell is determined by the interplay of the inward tug of gravity and the net effect of the kicks of the atoms that strike each face of the shell.

The calculation proceeds in time-steps of duration δt. At the start of each step, the speed and position of each atom are known, and the computer finds the position at a time δt later. This becomes the starting position for the next time-step, and the particles are assumed to move under the influence of constant gravity during each time-step. The shell is also accelerated inward by gravity (according to the inverse square of its distance from the center of the star) as well as being kicked outward by collisions.

The detailed result of each collision between an atom and a shell is found by requiring conservation of momentum and kinetic energy.

Each simulation starts with the shell at a prescribed radius and moving at a prescribed velocity. The atoms all move with the speed v_a, half of them inward and half outward. The program displays each atom and shell after each time-step and it plots the changing radius of the shell and the pressure of the gas.

Once the mass of the shell and the number of atoms are selected, the motion depends only on the starting values of the atom speeds and the shell motion. Suppose, as a special case, the shell is initially motionless. For each choice of atom speed, there is an equilibrium shell radius such that the shell will remain nearly motionless. This is the radius for which the inward and outward forces on the shell are in balance. The resulting plot is rather dull, showing a constant radius, with small fluctuations produced by individual kicks. This behavior corresponds to a static, nonpulsating star. On the other hand, if the shell is initially moving inward, it will overshoot the equilibrium, then rise up again and continue to oscillate indefinitely. In Figure 15.5, the radius and pressure are plotted as functions of time for a simulation consisting of two shells. Note

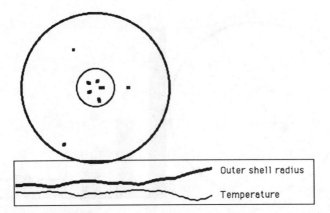

FIGURE 15.5 Simulation of star by two shells with 8 atoms moving radially and bouncing elastically from the shells. The radius and gas temperature are shown as functions of time (see text).

that the highest temperature (and pressure) corresponds to the smallest radius. It is the increase of pressure that accelerates the shell outward again.

15.4 STELLAR HEAT TRANSFER BY THE DIFFUSION OF PHOTONS

We now turn to another aspect of stellar interiors: the transfer of heat away from the site of nuclear burning by the random flight of photons. (Atoms do not carry much heat, because they are constantly colliding with each other and can move only a short distance between collisions.) Energetic photons are created in the high-temperature interior. They carry out random walks until they are absorbed and their energy heats the gas. A short time later, a new photon is formed (usually of lower energy) at the place where the original one was absorbed. It flies at random until it too is absorbed. This birth and death cycle is repeated until the photons arrive at the star's surface. From there, they escape into space, carrying away a bit of the star's energy.

In a real star, the photons take longer and longer steps as they move outward, but the most elementary simulation ignores this change of step-length and is based on a simple random walk in which each photon moves through a stationary matrix.

Figure 15.6 shows the results of a simulation in which photons are created at equal intervals in the center of a two-dimensional, disklike star and then execute random walks. The simulation keeps track of each photon and constructs a histogram showing the resulting density of photons at various distances from the center.

Figure 15.7 shows a variation on this theme. Here we see a disklike star in

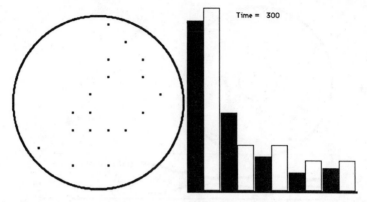

FIGURE 15.6 Photon diffusion in a disk-shaped star with a radius of 10 units. The star is divided into five rings of equal area and photons are injected at the center every 2 time-steps. They execute two-dimensional random walks of unit step-length along x and y, escape when they reach the surface. The histograms show the average (solid bars) and instantaneous (hollow bars) numbers of photons in each ring.

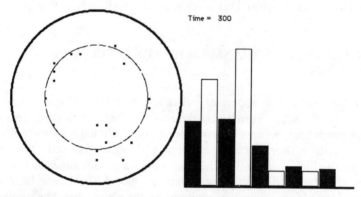

FIGURE 15.7 Similar to Figure 15.6 except the photons are injected at the interface between the second and third rings, at a radius of 6.3 units. The histograms show the average (solid bars) and instantaneous (hollow bars) numbers of photons in each ring. Note that the density of photons in the interior is constant, corresponding to the formation of a contant-temperature interior. Compared with the simulation in Figure 15.5, the photons spent less time inside the star.

which the photons are created at a finite and contant distance from the center of the star. This imitates the fact that, as a star ages, the original fuel in the interior is depleted, so the nuclear processes move outward. The result is a star that has a constant temperature inside the shell of burning. (This constant temperature of the interior may not be intuitively obvious, but it follows from energy conservation. In order for the center to be the hottest part of the star, it must have a source of energy to compensate for the energy carried away. This is familiar from heat-conduction problems.)

15.5   STELLAR PULSATIONS DAMPED BY HEAT FLOW

When we combine the two processes discussed earlier in this chapter (motion in gravitational field and heat transfer by diffusion), we find a remarkable behavior. The system will oscillate violently if the parameters of the system are properly adjusted. The motion can be quite regular or highly irregular and chaotic, depending on the relationship between the dynamic and the thermal properties.

This behavior can be considered as a simulation of pulsating variable stars, whose brightnesses are observed to change by factors of 2 to 10 over periods of days to years. Computations based on conventional hydrodynamic formulations have shown that a star can behave as a heat engine, undergoing oscillations that are driven by the heat that is escaping through the outer layers from the interior.

REFERENCES

Atkins 1984
Baker 1986
Feynman 1972
Prigogine 1978
Vincenti and Kruger 1965
Whitney 1984
Yourgrau 1982

PART FOUR

STOCHASTIC METHODS OF PARAMETER ESTIMATION

16

ERRORS OF MEASUREMENT

Up to this point, we have focused on the computer generation of numbers that simulate the behavior of stochastic physical systems. These systems were described by sets of parameters, such as energy or number of particles. In this chapter, we turn the problem around and discuss the problem of *deriving the values of parameters* from measured data in the presence of errors.

16.1   INTRODUCTION

Statistical methods of data analysis that were developed before the widespread availability of electronic computers were designed to reduce the amount of arithmetic required of the analyst. They usually started—and ended—with the assumption that the errors of measurement are normally distributed, that is, follow a gaussian distribution. This assumption implies that a single parameter, the variance of the measured values, predicts the expected errors of each size.

But not all measurements have errors that follow the normal distribution, so methods of greater flexibility are often required. The application of electronic computers and the recent development of stochastic techniques have permitted treating errors that follow a law that (1) is different from the normal distribution or (2) is unknown to the analyst.

The next three chapters describe methods of fitting a theoretical model to a set of observed data to find the best fit between the data and a theoretical model. The model is defined by a set of parameters, and our task is to evaluate the best values for the parameters by comparing the theory with the observations.

When the data are abundant and the errors of observation are independent of each other and obey the normal distribution, we can often use the "least squares" method to evaluate the parameters (Sections 16.2 and 17.1). When the errors follow an unknown law, we can resort to "bootstrapping" (Section 16.3). When the data are sparse or the errors follow a complicated law, we can use the maximum likelihood method (Sections 17.2–17.3). More general problems of optimum fitting are discussed in Chapter 18.

16.2 NORMALLY DISTRIBUTED ERRORS

(a) Biased and Unbiased Estimates of Measurement Error

Suppose we wish to determine the value of a quantity that can be assumed to have the constant value T during the process of measurement. Each measurement is subject to an error e_i, so if we repeat the measurement N times, we obtain distinct values $T_i = T + e_i$ $(i = 1, \ldots, N)$. We ask the following questions:

(1) What is the best estimator for the true value T?
(2) What is the best estimator for the root-mean-square uncertainty of a single measurement $T_i - T$?

As a first step, let us suppose that we know the true value T. Then an *unbiased estimator* of the mean squared error could be obtained from

$$
\begin{aligned}
s_\infty^2 &\equiv \langle e^2 \rangle \\
&= \frac{1}{N} \sum e_i^2 \\
&= \langle (T_i - T)^2 \rangle \\
&= \frac{1}{N} \sum (T_i - T)^2.
\end{aligned}
\tag{16.1}
$$

As an example of the process, the program GAUSS was used to simulate the measurement of a quantity whose true value is $T = 50$ and which is subject to a mean squared error of 5. (See discussion of GAUSS in Chapter 12.) Thus $M = 50$ and $D^2 = \langle e^2 \rangle = 5$. In 12 series of runs, three "measurements" were generated by calling GAUSS three times, with the results shown in columns 1–3 of Table 16.1a. The remainder of this table shows the calculation of the variances of the data. Column 4 shows the true value T. Columns 5–7 show the differences from the true value, and columns 8–10 show squares of these differences. Column 11 shows the variance of each series and the boldfaced figures at the bottom show the means over all series. Thus the overall variance of these simulated data was 4.66, not far from the expected value of 5. Individual

TABLE 16.1a Simulated Measurement Errors

	"Measured" Values			True	$T_i - T$			$(T_i - T)^2$			Mean
Column	1 T_1	2 T_2	3 T_3	4 T	5	6	7	8	9	10	11
Series											
1	50.22	51.91	53.88	50.00	0.22	1.91	3.88	0.05	3.65	15.05	6.25
2	48.99	49.03	45.00	50.00	-1.01	-0.97	-5.00	1.02	0.94	25.00	8.99
3	49.91	54.54	50.19	50.00	-0.09	4.54	0.19	0.01	20.61	0.04	6.89
4	49.27	52.11	54.13	50.00	-0.73	2.11	4.13	0.53	4.45	17.06	7.35
5	48.50	48.89	47.45	50.00	-1.50	-1.11	-2.55	2.25	1.23	6.50	3.33
6	50.94	49.92	50.40	50.00	0.94	-0.08	0.40	0.88	0.01	0.16	0.35
7	49.73	50.39	51.48	50.00	-0.27	0.39	1.48	0.07	0.15	2.19	0.81
8	53.62	50.40	48.95	50.00	-3.62	0.40	-1.05	13.10	0.16	1.10	4.79
9	45.75	51.66	52.06	50.00	-4.25	1.66	2.06	18.06	2.76	4.24	8.35
10	52.03	49.88	51.07	50.00	2.03	-0.12	1.07	4.12	0.01	1.14	1.76
11	48.41	47.16	49.72	50.00	-1.59	-2.84	-0.28	2.53	8.07	0.08	3.56
12	47.19	51.00	48.69	50.00	-2.81	1.00	-1.31	7.90	1.00	1.72	3.50
Mean	49.55	50.57	50.25					4.21	3.59	6.19	4.66

series of measurements, considered separately, give variances that fluctuate about this value.

Exercise 16.1
Using Table 16.1a, show that the fluctuations are smaller when the data are taken in groups of 6.

But, of course, in a typical situation we do not know the true value T, or else we would not be doing the measuring. We must work only with *observed* quantities. A reasonable procedure for finding the best estimator of T would be to take the observed mean $\langle T_i \rangle$. (In Chapter 17 we shall show that this choice is, in fact, the most likely to be correct if the errors are normally distributed.) Having chosen $\langle T_i \rangle$ as the estimator, we could estimate the error of each measurement from $e'_i = T_i - \langle T_i \rangle$. We use the prime to distinguish this error estimate from the true (but unknown) error, $e_i = T_i - T$. By analogy with our earlier expression, we could then estimate the mean squared error for N measurements from

$$s_N^2 \equiv \frac{1}{N}\sum e_i'^2. \tag{16.2}$$

This is called the *biased estimator* of the error, because it is measured from the *mean of the sample* rather than from T, the true mean of the population from which the sample is drawn. The biased estimator will be smaller than the unbiased estimator. Thus s_N^2 underestimates the actual error. (Can you see why?) We can verify this from the data in Table 16.1b, which shows the evaluation of the biased estimator. The structure of this table is the same as that of Table 16.1a, except that column 4 contains the *observed mean* of each series, rather than the unknown *true value*.

Exercise 16.2
Use the data of Table 16.1b to show that the difference between the true mean and the observed mean (the bias error) is relatively small when $N = 12$. Do this by regrouping the data into three series, forming them from each of the columns 1–3. Show that the biased estimator is $s_N^2 = 4.45$ in this case.

Let us now look at the theoretical relation between s_∞^2 and s_N^2. This will permit us to compute an unbiased estimator of measurement error from the observed data themselves. We start from

$$T_i - \langle T_i \rangle = T + e_i - \left(T + \frac{1}{N}\sum e_i \right)$$

$$= e_i - \frac{1}{N}\sum e_i. \tag{16.3}$$

TABLE 16.1b Simulated Measurement Errors

Column Series	"Measured" Values			Mean	$T_i - \langle T_i \rangle$			$(T_i - \langle T_i \rangle)^2$			Mean
	1 T_1	2 T_2	3 T_3	4 $\langle T_i \rangle$	5	6	7	8	9	10	11
1	50.52	51.91	53.88	52.00	-1.78	-0.09	1.88	3.18	0.01	3.52	2.24
2	48.99	49.03	45.00	47.67	1.32	1.36	-2.67	1.73	1.84	7.15	3.57
3	49.91	54.54	50.19	51.55	-1.64	2.99	-1.36	2.68	8.96	1.84	4.49
4	49.27	52.11	54.13	51.84	-2.57	0.27	2.29	6.59	0.07	5.26	3.97
5	48.50	48.89	47.45	48.28	0.22	0.61	-0.83	0.05	0.37	0.69	0.37
6	50.94	49.92	50.40	50.42	0.52	-0.50	-0.02	0.27	0.25	0.00	0.17
7	49.73	50.39	51.48	50.53	-0.80	-0.14	0.95	0.65	0.02	0.90	0.52
8	53.62	50.40	48.95	50.99	2.63	-0.59	-2.04	6.92	0.35	4.16	3.81
9	45.75	51.66	52.06	49.82	-4.07	1.84	2.24	16.59	3.37	5.00	8.32
10	52.03	49.88	51.07	50.99	1.04	-1.11	0.08	1.07	1.24	0.01	0.77
11	48.41	47.16	49.72	48.43	-0.02	-1.27	1.29	0.00	1.61	1.66	1.09
12	47.19	51.00	48.69	48.96	-1.77	2.04	-0.27	3.13	4.16	0.07	2.46
Mean	49.55	50.57	50.25					3.57	1.86	2.52	2.65

Next, we square both sides,

$$(T_i - \langle T_i \rangle)^2 = e_i^2 - \frac{2e_i}{N}\sum e_i + \frac{1}{N^2}\sum e_i \sum e_i.$$

We then sum over N, finding

$$\sum (T_i - \langle T_i \rangle)^2 = \sum e_i^2 - \frac{2}{N}\sum e_i \sum e_i + \frac{1}{N}\sum e_i \sum e_i$$

$$= \sum e_i^2 - \frac{1}{N}\sum e_i \sum e_i. \tag{16.4}$$

For simplicity, we suppose that the errors all are the same size but with randomly selected signs, so $e_i = \pm e$. And we introduce the random factor $f_i = \pm 1$, so $e_i = f_i e$. Then the average error is

$$\langle \sum e_i \rangle = \langle (f_1 + f_2 + \cdots) \rangle e = 0. \tag{16.5a}$$

and the average of the first sum on the right of (16.4) is

$$\langle \sum e_i^2 \rangle = Ne^2. \tag{16.5b}$$

We can evaluate the product of sums in (16.4). Now we can write for the average values

$$\sum e_i \sum e_i = (f_1 + f_2 + \cdots + f_N)(f_1 + f_2 + \cdots + f_N)e^2. \tag{16.6}$$

The product on the right has two types of term:

$$f_i f_i = 1$$

and

$$f_i f_i = \pm 1, \quad \text{when } i \neq j.$$

The second type of term will mutually cancel when averaged over a large number of samples. Thus

$$(f_1 + f_2 + \cdots)(f_1 + f_2 + \cdots) = \sum f_i^2 = N, \tag{16.7}$$

and (16.6) becomes

$$\sum e_i \sum e_i = N \langle e^2 \rangle. \tag{16.8}$$

(Note the similarity of this relation to the mean squared displacement of a random walk, where N is the number of steps of length e.) Now we have from (16.4), (16.5b), and (16.6) that

$$\sum(T_i - \langle T_i \rangle)^2 = N \langle e^2 \rangle - \langle e^2 \rangle.$$

dividing by N and comparing with (16.2), we have the desired relation for the biased estimator in terms of the true errors,

$$s_N^2 = \frac{(N-1)\langle e^2 \rangle}{N}. \qquad (16.9)$$

Comparing with (16.1), $s_\infty^2 = \langle e^2 \rangle$, we see that the biased and unbiased estimates of the mean squared error are related by

$$s_N^2 = \frac{(N-1)}{N} s_\infty^2. \qquad (16.10)$$

Thus the biased estimator is smaller than the unbiased estimator, but when $N \gg 1$, the ratio of the two estimators is nearly unity and the distinction is unimportant.

Exercise 16.3
Verify that (16.10) is a good approximation for a finite sample of data using Table 16.1b.

(b) Estimating the Error of a Mean Value

We now ask what the uncertainty of our estimator of T is. That is, what uncertainty is to be attached to the estimator of the mean, and how does this uncertainty depend on N, the number of measurements?

To avoid a possible source of confusion, it is important to distinguish between the *number of runs* (12 in Table 16.1b) and the *number of measures* in each run ($N = 3$). The scatter of the means does not depend on the number of runs, but only on the number of measures in each run and on the error of individual measures.

Suppose we have N measurements, and the estimated error of a single measure is e. Let s_μ^2 be the mean squared difference between the sample mean and the true mean. It is evaluated from a random walk of N steps of length e. That is,

$$s_\mu^2 \equiv \langle (\langle T_i \rangle - T)^2 \rangle = \frac{e^2}{N}. \qquad (16.11)$$

This is often called the *variance of the mean.*

Exercise 16.4
Derive (16.11) for s_μ^2 by starting from the expression $T_i - T = e_i$, summing over N observations, dividing by N and squaring the resulting expression.

Exercise 16.5
Use the data of Table 16.1b to verify that the variance of the mean decreases in proportion to the number of measures, as predicted by (16.11).

(c) Combined Errors

Suppose that instead of measuring a single quantity we measure two (such as the noon and midnight temperatures on a particular day), and we wish to evaluate the *variance of the difference*. The distribution of measurement errors of the noon temperature can be treated as a random walk with a certain mean squared step-length, and the same is true of the midnight temperature. The error of the difference of temperatures can be considered as a series of two random walks. The second walk starts where the first one ended.

In Exercise 5.1, we saw that the mean squared displacement of a combined walk is the sum of the mean squared displacements of the component walks. So, if the noon and midnight temperatures, T_1 and T_2, have mean squared errors e_1^2 and e_2^2, respectively, and the mean $\langle T \rangle_1$ is derived from N_1 measures while $\langle T \rangle_2$ is derived from N_2 measures, the variance of each mean will be e_1^2/N_1 and e_2^2/N_2, respectively. The variance of the difference (or the sum) is the sum of the individual variances,

$$s_{\text{diff}}^2 = s_{\text{sum}}^2 = \frac{e_1^2}{N_1} + \frac{e_2^2}{N_2}. \tag{16.12}$$

Exercise 16.6
The following table gives fictitious measures of T_1 and T_2 (below). Derive the biased and unbiased estimates of the mean squared errors, if individual measures e_1^2 are 0.07 and 0.084 and e_2^2 are 0.06 and 0.08. Show that the unbiased variances of the means are 0.014 and 0.013, respectively. Thus the variance of the difference of the means $\langle T_1 \rangle - \langle T_2 \rangle = 0.027$, corresponding to 0.16 degrees.

Noon and Midnight Temperature Measurements
Noon, T_1 78.1, 78.9, 78.3, 78.6, 78.7, 78.4
Midnight, T_2 50.3, 50.6, 50.1, 50.4, 50.7, 50.3

(d) Confidence Limits Using the Probability Integral

If the error of a measured quantity x follows the normal distribution with a variance s^2, then the probability that the measured value will deviate by $x - \langle x \rangle$ from the mean is given by the expression (cf. Chapter 12)

$$p(x - \langle x \rangle)dx = \sqrt{2/\pi s^2}\, \exp\left(-\frac{(x - \langle x \rangle)^2}{2s^2}\right)dx. \tag{16.13}$$

The value of this expression is symmetric about the mean (because the square of $x - \langle x \rangle$ is the argument and all other terms are constant), so we can use the absolute value of the distance from the mean, in units of the standard deviation s, as a new variable:

$$r \equiv \frac{|x - \langle x \rangle|}{s}.$$

If we know the variance, we can estimate the likelihood that a measured value will lie between 0 and the value r' by evaluating the probability integral,

$$P(0 < r < r') = \int_0^{r'} p(r)\, dr$$

$$= 2 \sqrt{\frac{2}{\pi}} \int_0^{r'} \exp\left(\frac{-r^2}{2}\right) dr. \tag{16.14}$$

As r must have some value in the range $0 < r < \infty$, we have the normalization

$$P(0 < r' < \infty) = \int_0^{\infty} p(r)\, dr = 1. \tag{16.15}$$

Table 16.2 and Figure 16.1 show values of the probability integral. The displacement r is measured in units of the square root of the variance s, so the argument of the table is $r = x/s$. (There is no closed expression for its value, so we usually resort to a table of this type.)

From this table we infer, as an example, that the measured value will lie within $1.5s$ of the mean in 97% of the cases. Or, to put it another way, we can say with 97% confidence that the measured value will lie between $\langle x \rangle - 1.5s$

TABLE 16.2 Probability Integral

r	$p(0 < r' < r)$	r	$p(0 < r' < r)$
0.000	0.000	1.000	0.843
0.100	0.114	1.100	0.880
0.200	0.223	1.200	0.910
0.300	0.329	1.300	0.934
0.400	0.428	1.400	0.952
0.500	0.520	1.500	0.966
0.600	0.603	1.600	0.976
0.700	0.678	1.700	0.984
0.800	0.742	1.800	0.989
0.900	0.797	1.900	0.993

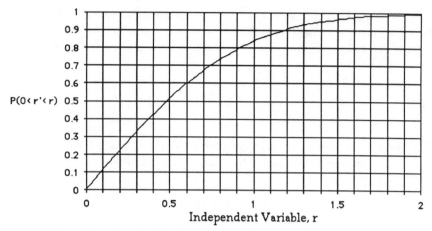

FIGURE 16.1   Plot of probability integral from Table 16.2. The independent variable is $r = x/s$, where x is the measured value and s is the variance of the errors. If the errors truly follow the normal distribution, this shows the probability that an error will be less than x. For example, 90% of the errors will be less than $1.2s$.

and $\langle x \rangle + 1.5s$. In this way, the gaussian (normal) model for error distribution permits us to put confidence limits on expectations.

Exercise 16.7
On a common y scale, plot a graph of $p(y)$, evaluated at a convenient set of points, and the probability integral $P(y)$ in the table. Show that the slope of the probability integral at y is proportional to $p(y)$.

16.3   BOOTSTRAPPING WITH AN UNKNOWN ERROR DISTRIBUTION

Suppose we do not know the distribution of errors and we suspect that it does not follow the normal curve. All is not lost. We can still establish confidence limits from a set of data, using a method that is computer intensive and would not have been practical in the days before electronic computers. It is called "bootstrapping" because it gives the appearance of lifting oneself by one's own bootstraps. Despite its name, it does not perform miracles, but it works with a minimum of assumptions and does not exclude the possibility that the errors are normally distributed. See Kinsella (1986) for a brief discussion and illustration of this method and a related method known as "the jack knife."

Suppose we have seven measures of the length of a table, as listed below:

Measured Lengths (inches)

i	1	2	3	4	5	6	7
Value	16.1	16.1	16.3	16.4	16.4	16.6	16.6

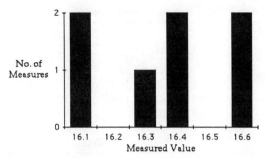

FIGURE 16.2   Frequency distribution of seven hypothetical measures of the length of a table. The errors of these measures may not follow a normal distribution, so we use bootstrapping to find the best estimate and its uncertainty. See Figure 16.3.

Exercise 16.8
Show that the straight mean of these values of 16.32 and the standard deviation is 0.21. From the probability integral and the assumption that these values obey the normal distribution, verify that, with a confidence level of 95.2%, we can say that the values lie in the range 16.32 ± 0.29 or 16.03 to 16.61.

These data (Figure 16.2) do not follow a normal distribution, so we require a more flexible method of estimating the best value and setting confidence limits.

(a) Sampling with Replacement

Let us draw seven samples (with replacement) from this list of measures by randomly picking seven integers in the range $1 < i < 7$ and taking the corresponding value from the table. The same value can be picked more than once, hence the name "with replacement," because it is as though we had put each number back into a bag after having drawn it.

We draw sets of seven samples a large number of times, say, $N = 250$, evaluating the mean $L_j (j = 1 - N)$ for each set. We then prepare a histogram

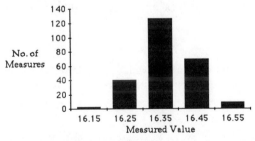

FIGURE 16.3   Result of repeated sampling of the frequency distribution in Figure 16.2. As described in the text, this "bootstrapping" is useful when the errors do not follow a normal distribution. It provides a hypothetical set of measures whose mean value and spread will give an estimate of the data mean and uncertainty.

TABLE 16.3   Results of Bootstrapping

Bin	Range	Mean	Frequency	%
1	16.05–16.15	16.1	3	1.2
2	16.15–16.25	16.2	41	16.4
3	16.25–16.35	16.3	127	50.8
4	16.35–16.45	16.4	70	28.0
5	16.45–16.55	16.5	9	3.6

Total 250

(see Figure 16.3 and Table 16.3), showing the frequency distribution of the derived *mean* values.

(b) Estimating the Mean and Variance

As an estimator of the mean, we take the overall mean of the derived means,

$$\langle L_j \rangle = \frac{1}{N} \sum_{j=1}^{N} L_j = 16.32.$$

We may derive an unbiased estimate of the uncertainty of a measurement, s, from the square root of the expression

$$s^2 = \frac{1}{N-1} \sum_{j=1}^{N} (L_j - \langle L_j \rangle)^2.$$

We may also derive confidence limits for various intervals from the data in Table 16.3. For example, we see that 95.2% of the values L_j fell in bins 2–4, corresponding to the interval 16.15–16.45. This would lead us to say with 95.2% confidence that the length lies in this interval. (They are not exactly symmetric about the mean because the errors are distributed in an asymmetric fashion.)

Thus the bootstrapping method assumes that the errors are well represented by the table of measured values, and it lets the tabulated frequencies speak for themselves.

Exercise 16.9

Construct a fictitious set of measures with a more lopsided distribution of errors and contrast the results of bootstrapping with the results of assuming a normal distribution of errors.

REFERENCES

Efron and Morris 1977
Kinsella 1986

17

ESTIMATION OF MULTIPLE
PARAMETERS

In the previous chapter, we concentrated on estimating the value of a constant from measured data. Now we look at the more general problem of deriving several parameters $\{a_j\}$ from a set of data. For example, suppose we observe a ball move across a table and we measure the ball's position y_i at a sequence of times t_i $(i = 1, 2, \ldots, N)$. These measurements may be considered as a set of N events, each characterized by a value of the independent variable t. Furthermore, suppose we have a theory for the motion of this ball that will predict its position at various times. This theory involves the initial position and the initial speed of the ball, as well as its acceleration. These are the parameters $\{a_j\}$ of the theory. On the basis of the observations, we wish to assign numerical values to the parameters.

We may proceed as follows. From the theory and a model for the errors of measurement, we can compute the relative probability of each measured value. Call this probability $P(y_i, t_i, \{a_j\})$, where t_i is the time of measurement. If the measurements are statistically independent events (as described in Appendix 1), the entire data set has a joint probability given by the product

$$P(y_1, t_1, \{a_j\})P(y_2, t_2, \{a_j\}) \cdots P(y_N, t_N, \{a_j\}) = \prod_{i=1}^{N} P(y_i, t_i, \{a_j\}). \quad (17.1)$$

We then look for the values of the set of parameters $\{a_j\}$ for which this joint probability is maximum. We adopt these parameters as the best fit of the theory to the observations, because they are the values for which the observed data have the greatest likelihood. These are the values that are most likely to be implied by the data.

17.1 ABUNDANT DATA AND NORMALLY DISTRIBUTED ERRORS

(a) Linear Least Squares

As an example, we might expect the distance of the ball, η, to increase with time according to the simple linear relation $\eta = a + vt$, if it started from point $\eta = a$ with constant speed v. We take measurements y_i of the distance at a series of known times t_i, and each measurement is subject to a random error e_i, so

$$y_i = \eta_i + e_i = a + vt_i + e_i. \tag{17.2}$$

We wish to find the values of a and v that give the best fit between the observed set of measurements $\{y_i\}$ and the theoretical model. (See Figure 17.1.)

We assume that the errors are drawn from a population that does not change during the course of the experiment. Furthermore, we assume that the errors obey the normal distribution described in Chapter 13. In this case, e has the probability density function,

$$p(e) = \frac{1}{s\sqrt{\pi}} \exp\left(\frac{-e^2}{2s^2}\right), \tag{17.3}$$

in which s is the root-mean-square error obeying

$$s^2 \equiv \int_{-\infty}^{\infty} e^2 p(e)\, de. \tag{17.4}$$

The numerical value of s is usually not known in advance. It is to be derived from the measurements.

If we assume that a precise measurement at time t_i would give the value

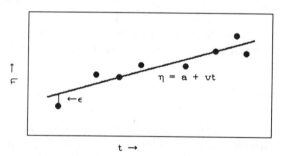

FIGURE 17.1   Measurements subject to a random error e are made on a quantity η, which is expected to obey a linear relation. Each measurement is indicated by a dot, and our task is to derive the "best" values for a and v.

$\eta_i = a + vt_i$, the probability that the measured value at time t_i will lie in the range $(y, y + dy)$ is then determined by the probability of the corresponding error $e = y - \eta$, or

$$P(y_i)\, dy = p(e_i)\, de = \frac{1}{s\sqrt{\pi}} \exp\left(\frac{-(y_i - \eta_i)^2}{2s^2}\right) dy, \qquad (17.5)$$

where

$$\eta_i = a + vt_i.$$

The probabilities depend on the difference from the theoretical value η_i, so they are functions of the parameters a and v.

If the data consist of a set of N statistically independent observations, giving values $y_i \{i = 1, \ldots, N\}$, the joint probability of observing exactly this set of values is the product $\prod P(y_i)\, dy_i$ of the individual probabilities. This joint probability, like the individual probabilities, is a function of the adjustable parameters a and v.

The final step is to find those values of a and v that maximize this probability. Instead of maximizing $\prod P(y_i)\, dy_i$ it is more convenient (and amounts to the same thing) to *minimize* the cost function C, defined by

$$C \equiv -\ln\{\prod P(y_i)\} \quad \text{to be minimized.} \qquad (17.6)$$

The values of a and v that make this function a minimum will also make $\prod P(y_i)$ a maximum. Note that the dy_i have dropped out because they are independent of the parameters a and v. Since the logarithm of the product is the sum of the logarithms, we may write the cost function as

$$C = -\ln\{P(y_1)\} - \ln\{P(y_2)\} - \cdots$$
$$= g \sum_{i=1}^{N} (y_i - \eta_i)^2,$$

where N is the total number of data points and g is a constant of proportionality which we may set equal to unity. So we seek the minimum of the expression

$$C = \sum (y_i - \eta_i)^2 = \sum (y_i - a - vt_i)^2 \quad \text{to be minimized.} \qquad (17.7)$$

The "least squares" method gets its name from this expression, which contains a sum of squares. Remember that this expression is a result of our having *assumed a normal distribution for the errors.*

Equations for the values of a and v corresponding to the minimum of C can be found by elementary calculus. We set the derivatives of C as given by (17.7) with respect to a and v individually equal to zero, and this leads to two equations in a and v. From $d\{C\}/da = 0$, we find the first equation,

$$\sum (y_i - a - vt_i) = 0, \qquad (17.8a)$$

and from $d\{C\}/dv = 0$, we find the second equation,

$$\sum (y_i t_i - at_i - vt_i^2) = 0. \qquad (17.8b)$$

These may be rewritten as the "normal equations,"

$$Na - v\sum t_i = \sum y_i, \qquad (17.9a)$$

$$a\sum t_i + v\sum t_i^2 = \sum y_i t_i. \qquad (17.9b)$$

We evaluate the sums and find a and v from these two simultaneous linear equations. (A little further analysis permits finding s as well as estimates of the uncertainties of a and v.)

Table 17.1 shows a sample set of seven observations that can be analyzed by this method. The first two columns give the time and the measured positions. These are the data, and we fit them with the linear expression

$$y_i = a + vt_i + e_i.$$

By performing the indicated summations, the normal equations are found to be

$$7a + 39v = 85.5,$$

$$39a + 339v = 720.0,$$

whose solution is

$$a = 1.0616, \qquad v = 2.0018.$$

TABLE 17.1   Sample Data and Computation of Normal Equations

i	t	y	t^2	yt	η	e	e^2
1	0	0.5	0.0	0.0	1.062	−0.562	0.315
2	1	4.0	1.0	4.0	3.063	0.937	0.877
3	3	7.5	9.0	22.5	7.067	0.433	0.188
4	6	12.0	36.0	72.0	13.072	−1.072	1.150
5	7	14.5	49.0	91.5	15.074	−0.574	0.329
6	10	22.0	100.0	220.0	21.079	0.921	0.848
7	12	25.0	144.0	300.0	25.083	−0.083	0.007
Sums	39	85.5	339.0	710.5	—	0.000	3.71

These are the "best" parameters in the least squares sense. Using them, we predict values for η in the sixth column, and from these and the data of the third column we evaluate the individual errors of measurement, indicated in the seventh column. The unbiased root-mean-squared error is found to be

$$s = \sqrt{\frac{1}{N-1}\sum e^2} = 0.79. \qquad (17.10)$$

The solution to this problem is readily executed with a spreadsheet, as shown in Appendix 3.

Exercise 17.1
In the absence of air resistance, a body near the Earth's surface falls with constant acceleration according to the formula $s = vt - gt^2/2$, where s is the height above the starting point, v is the initial upward velocity, and g is the downward acceleration of gravity. Suppose the following data were obtained from measurements at five times:

Time (seconds)	Distance (meters)	Time (seconds)	Distance (meters)
1	5	4	-38
1.5	4	5	-72
3	-14		

Show that the function to be minimized is

$$C = \sum \left(s - vt + \frac{gt^2}{2} \right)^2$$

and that the normal equations are

$$v \sum t^2 - \frac{g}{2}\sum t^3 = \sum st,$$

$$v \sum t^3 - \frac{g}{2}\sum t^4 = \sum st^2.$$

Use a spreadsheet program to show that these lead to the values $v = 9.948$ and $g = 9.736$, and the root-mean-square error of a measurement is 0.062 meters.

(b) Reduction of Nonlinear Problem to Linear Least Squares

The linearity of (17.9) and the equations for the parameters in Exercise 17.1 are a consequence of the facts that the theoretical equation $s = vt - gt^2/2$ is linear in the parameters v and g (i.e., they do not appear in a polynomial or transcendental function) and that we assumed the errors to follow the normal distribution.

This linearity is a great help in finding a solution. Often a problem looks nonlinear and difficult to solve, but it can be turned into a linear problem by a change of variables. We illustrate this using a different physical phenomenon, radioactive decay (Cleveland 1983).

Suppose we have a sample of a radioactive element and we measure times of individual decays with a Geiger counter (Figure 17.2). Our task is to determine the rate of the decays and thereby to estimate the amount of radioactivity present. The expected rate of decays from the sample will decrease exponentially with the probability density $a \exp(-\lambda t)$. In this expression, $\lambda = 1/\text{(half-life)}$. The multiplying constant a depends on the quantity of sample being counted, and the half-life depends on the particular chemical element that is decaying. We wish to determine a and λ from the data.

The data consist merely in the times of the individual decays, and in order to set up a relationship between a dependent and an independent variable, we divide the total observation time into intervals Δ and let n_i be the observed number of counts in the ith interval. In this way, we obtain the independent variable t_i and the corresponding dependent variable n_i.

The probability of an event in the interval dt is

$$\rho(t) \, dt = a \exp(-\lambda t) \, dt. \tag{17.11}$$

The expected number of events, $\langle n_i \rangle$, during the interval T_i to $T_{i+\Delta}$ is the integral

$$\langle n_i \rangle = \int_{T_i}^{T_i + \Delta} \rho(t) \, dt, \tag{17.12}$$

which gives

$$\langle n_i \rangle = c \exp(-\lambda T_i), \tag{17.13}$$

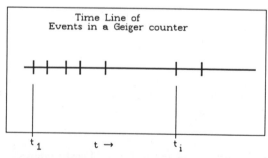

FIGURE 17.2 Schematic time-line for events triggered in a Geiger counter by radioactive decays. The time of each event is assumed to be known precisely. Our task is to fit the best exponential probability curve to the occurrences of events. In the least squares procedure, we first divide the time-line into intervals and count the events in each interval.

where we define the constant

$$c \equiv a[1 - \exp(-\lambda\Delta)].$$

Our task is to find values of a and λ that best fit the measured values n_i. The parameter λ is involved in a transcendental function, so the problem is nonlinear, but it can be reduced to a linear problem. Taking logarithms of both sides of this expression we have

$$\ln\langle n_i \rangle = \ln c - \lambda T_i. \tag{17.14a}$$

This can be represented by the linear equation

$$y_i = y_0 - \lambda T_i, \tag{17.14b}$$

where we have introduced the new variables

$$y_i \equiv \ln\langle n_i \rangle, \tag{17.15a}$$

$$y_0 \equiv \ln c. \tag{17.15b}$$

The cost function to be minimized is

$$C = \sum (y_i - y_0 + \lambda T_i)^2, \tag{17.16}$$

and the normal equations may be found by equating to zero the derivatives of C with respect to y_0 and λ (see Exercise 17.2). After finding the parameters y_0 and λ, we may return to the original variables with (17.13) and (17.15b) written

$$c = \exp(y_0), \qquad a = \frac{c}{1 - \exp(-\lambda\Delta)}. \tag{17.17}$$

Exercise 17.2
Show that the normal equations are

$$y_0 N - \lambda \sum T_i = \sum y_i,$$

$$y_0 \sum T_i - \lambda \sum T_i^2 = \sum y_i T_i,$$

TABLE 17.2   Sample Count Data, $\Delta = 5$ seconds

i	T_i	n_i	$\ln(n_i)$	$\langle n_i \rangle$
1	0	357	5.878	379
2	5	180	5.193	175
3	10	80	4.382	81
4	15	45	3.807	37
5	20	15	2.708	17

and using the data of Table 17.2 evaluate y_0 and λ. Then find c and a from (17.17). (*Hint:* Start by evaluating $y_i = \ln(n_i)$ for each of the 5-second time intervals; then compute the necessary sums and insert them into the normal equations.) Verify the approximate values $\lambda = 0.1545$ and $a = 705$, and using these parameters, verify the values predicted by the theory, $\langle n_i \rangle$, shown in Table 17.2.

17.2   FAILURE OF THE LEAST SQUARES METHOD FOR SPARSE DATA

A key step of the previous section was to divide the time-line into intervals and evaluate the number of counts per interval. We did this in order to derive a function of time from the counts. This procedure works quite well as long as we have many counts in each time interval. We can take the numbers of counts as the dependent variable while the time associated with each interval is the independent variable.

However, when we have a small number of events, the division into finite time intervals leads to very "lumpy" data. When the actual times of the individual events are replaced by the mean time assigned to each interval, the data can become distorted. For example, if a finite time interval contains only two events, the mean time of those two events might differ by as much as $\Delta/2$ from the time at the center of the interval to which they are assigned.

17.3   MAXIMUM LIKELIHOOD FOR SPARSE DATA

For sparse data, we need a method that uses the actually measured individual times t_i, rather than the counts in finite intervals. We now describe such a method. This discussion is based on B. T. Cleveland's article (1983).

The data consist of the times of the events, t_i, and we start by computing the probability that the first event occurs at time t_1. As we showed in Chapter 11, if the probability of no decay in the infinitesimal interval dt is $1 - \rho(t)dt$, then the probability of no decay between $t = 0$ and $t = t_1$ is

$$\exp\left[-\int_0^{t_1} \rho(t)\, dt \right].$$

Therefore the probability that the first event will occur in the interval $(t_1, t_1 + dt_1)$ is the joint probability

$$P(t_1)\, dt = \exp\left[-\int_0^{t_1} \rho(t)\, dt \right]\rho(t_1)\, dt_1. \tag{17.18}$$

This is the product of two terms: The first is the probability that no event occurred before t_1, and the second, $\rho(t_1)\, dt_1$, is the probability that exactly one event did occur in the immediately adjacent interval dt_1. (We neglect the probability that two events occur in the short interval dt.)

In a similar manner, the probability that the second event will occur immediately after the observed interval $t_2 - t_1$ is given by

$$P(t_2) dt = \exp\left[-\int_{t_1}^{t_2} \rho(t) dt \right] \rho(t_2) dt_2. \qquad (17.19)$$

The joint probability that the events occurred exactly at the times $\{t_i\}$ is the product

$$\prod P(t_i) dt_i = \exp\left[-\int_0^{t_1} \rho(t) dt \right] \exp\left[-\int_{t_1}^{t_2} \rho(t) dt \right] \cdots \rho(t_1) dt_1 \, \rho(t_2) dt_2 \cdots. \qquad (17.20)$$

If T is the time of the last event, we can simplify this by noting that the product of exponentials can be replaced by a single expression,

$$\exp\left[-\int_0^t \rho(t) dt \right] \exp\left[-\int_{t_1}^{t_2} \rho(t) dt \right] \cdots = \exp\left[-\int_0^T \rho(t) dt \right].$$

So

$$\prod P(t_i) dt_i = \exp(-M) \prod \rho(t_i) dt_i, \qquad (17.21)$$

where we have defined

$$M \equiv \int_0^T \rho(t) dt.$$

We wish to maximize the probability, and we may accomplish this as before by taking its negative logarithm as the cost function to be minimized. This gives

$$C = -\ln\left[\exp(-M) \prod \rho(t_i) \right] = M - \sum \ln [\rho(t_i)], \qquad (17.22)$$

where the intervals dt_i have been removed because they are not adjustable.

From the expression for $\rho(t)$, we have $M = a(1 - e^{-\lambda T})/\lambda$. For the cost function, we have

$$C(a, \lambda) = \frac{a(1 - e^{-\lambda T})}{\lambda} - N \ln a + \lambda \sum t_i. \qquad (17.23)$$

We seek the values of a and λ that make this a minimum, and there are two general methods that we might use. First, we might try a blind search among values of a and λ, hoping to stumble on the pair that gives the smallest value of C. Second, we might be more analytical and start by differentiating C with respect to a and λ and setting the derivatives to zero. From the resulting expressions, we can solve for a and λ.

Rather than try a blind search, let us try the second method. We differentiate C in (17.23) with respect to a and set the result to zero, which gives

$$a = \frac{N\lambda}{1 - e^{-\lambda T}}. \tag{17.24}$$

Differentiating C in (17.23) with respect to λ and setting the result to zero gives, after some rearrangement,

$$\lambda = \frac{N[1 - \lambda T/(e^{\lambda T} - 1)]}{\sum t_i}. \tag{17.25}$$

Our task is to solve (17.24) and (17.25) for a and λ. Equation (17.25) contains only λ and known quantities, but we cannot solve it directly because it contains the exponential. We may, however, solve by trial and error for λ, and when this has been done, (17.24) may be solved for a. This completes the solution.

In general, the maximum likelihood method described in this section leads to a set of equations that must be solved iteratively. Whether we choose to do a blind search for the best values of a and λ or use the analytical method, the equations cannot be solved directly. We must search for the solution, and this is the subject of Chapter 18.

REFERENCE

Cleveland 1983

18

OPTIMIZATION

18.1 THE TRAVELING SALESMAN PROBLEM

We are frequently faced with a problem like that posed at the end of Chapter 17: a complicated function of N variables is to be minimized. Perhaps the most famous optimization problem is to find the shortest closed, nonrepeating route among N cities, as illustrated in Figure 18.1a. This is the traveling salesman problem, and the quality of a solution is judged by the numerical value of the length of the route.

Many problems that have no such simple geometrical interpretation can nevertheless be expressed as the search for the minimum of a cost function. In order to solve such problems, we must be able to define a set of parameters $\{a_i\}$ that prescribe a possible solution, and we must be able to evaluate a single number, $C(\{a_i\})$, which is a function of these parameters and which describes the quality of the solution. In the case of the traveling salesman problem, the set $\{a_i\}$ would be a list of cities in order of visit. From this list and the known coordinates of each city, we evaluate the total length of the route—or the square of the length. If the cities lie in a plane, we may take the cartesian distances:

$$C = \sum_{i=1}^{N-1} \sqrt{(x_i - x_{i+1})^2 + (y_i - y_{i+1})^2}.$$

This length is the "cost function" or the objective function we wish to minimize. Figures 18.1b and 18.1c illustrate two solutions for the map of Figure 18.1a.

The search through the values of $\{a_i\}$ for the set that gives the lowest cost function is very similar to the search for the deepest place in a lake bottom.

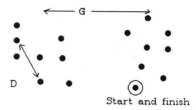

FIGURE 18.1a Schematic map of 16 cities (indicated by dots) illustrating the traveling salesman problem. The salesman wants to find the shortest route and avoid visiting any city more than once. In this map, the cities are divided into two groups separated by a distance G. Within each group, cities are typically separated by a distance D, where $D < G$.

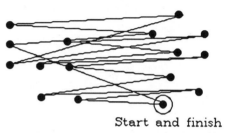

FIGURE 18.1b An obviously inferior solution requiring the traveler to go from one group to the other between every pair of visits. Contrast with Figure 18.1c.

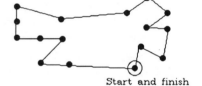

FIGURE 18.1c Improved solution in which traveler completes one group before moving to the next.

We start from an arbitrary place, perhaps near the middle of the lake, and then we move around, either following the local slope downward or moving at random. Whatever method we use, it is important not to mistake a local depression of the bottom for the deepest part of the lake.

18.2 OPTIMIZATION BY SIMULATED ANNEALING

In this section, we illustrate a stochastic method for finding the minimum of the cost function. The method was first used by Metropolis et al. (1953) for finding the minimum-energy configuration of a solid and has been described by Kirkpatrick et al. (1983). It has become a popular method in Monte Carlo simulations of many types.

We shall apply the method to the maximum likelihood problem of Chapter 17. The cost function $C(a, \lambda)$ is a surface whose local height depends on the two coordinates a and λ. Our goal is to find the lowest point of the surface, taking into account the fact that there may be many local minima.

We proceed by successive corrections. Starting from guessed values of a and λ we evaluate C from the expression in Chapter 17, namely,

$$C(a, \lambda) = \frac{a(1 - e^{-\lambda T})}{\lambda} - N \ln a + \lambda \sum t_i.$$

Then we randomly alter the parameters a and λ and we reevaluate the cost function, finding C' and the change $\delta C = C' - C$. If $\delta C < 0$, we accept the new parameters as defining a new starting point. If $\delta C > 0$, we reject or accept the new value on the basis of a probabilistic calculation. That is, we draw a random number, and we reject the new parameters if this number satisfies a certain constraint.

The probabilistic rejection is at the heart of the method; it permits the value of C to increase slightly, and this prevents the solution from getting stuck in local minima before it has reached the neighborhood of the absolute minimum that we seek. A convenient criterion for rejection is to compute $\exp(-\delta C/T)$, where T is an adjustable parameter that is given a fairly large value at the start of the Monte Carlo search. This formula is permissive of small increases of C but rarely permits a large increase.

At each step of the search, we first compute $\exp(-\delta C/T)$ and then we draw a random number r that is uniformly distributed in the range 0–1. If $r < \exp(-\delta C/T)$, we accept the new point (a', λ'), otherwise we reject it. When T is large, this formula permits moderately large increases of C. As the solution progresses and moves into the region of the absolute minimum, the rejection criterion is adjusted to make increases of C less and less likely. This is accomplished by reducing the value of T, making the likelihood of accepting a change for which $\delta C > 0$ smaller. Establishing the precise schedule for reducing T is a matter of experience; if T is reduced too quickly, the process may get stuck at a local minimum, corresponding to a high value of the cost function; if it is reduced too slowly, the process will wander indefinitely.

One possible schedule is illustrated below:

```
Procedure schedule;
{Keep track of progress and change temperature}
        Begin
        If nacceptances > totalQ Then
                Begin
                endFlag := endflag - 1;
                needTemp := true;
                End;
        If ctrial >= 2 * totalQ Then
                Begin
                endFlag := endFlag + 1;
```

```
                    needTemp := true;
                End;
            IF endFlag > 4 Then
                    endCond := true;
            ShowStatus;
            End;
```

Exercise 18.1

Find a solution to the traveling salesman problem for 12 cities arranged in two rectangular arrays of 6 cities. Start with an arbitrary route and before each alteration pick two cities at random and reverse the section of the route lying between them.

18.3   ANOTHER EXAMPLE: THE BOLTZMANN DISTRIBUTION

In Chapters 7 and 8 we saw that on a board with N squares the expected distribution function n_q for the number of squares with q chips (given the constraint that the total number of chips $Q = \sum q n_q$ is fixed) approached a geometric series for large systems. This function is the one that gives the greatest number of configurations, Γ, and hence corresponds to the greatest likelihood.

We now search for Γ using the method of random searching described in Section 18.2. The expression for Γ is (from Chapter 7)

$$\Gamma = \frac{N!}{n_0! n_1! \cdots n_Q!},$$

and we seek the set $\{n_q\}$ that leads to the maximum value, subject to the energy constraint $Q = \sum q n_q$. Because N is a constant and merely scales the value of Γ, we need not include it in the cost function, so we take the logarithm of the denominator,

$$C = \sum_{q=0}^{Q} \ln n_q!,$$

as the function to be minimized. We start from an arbitrary list of n_q $(q = 0, 1, \ldots, Q)$ consistent with the energy constraint, and by successive trials we attempt to alter this list so it produces the minimum value of C, which implies a maximum for Γ. The alteration of the list at each step (which corresponds to a step in Q-dimensional space) must leave Q unchanged, but we may use any method we choose. The final result is unaffected, although the speed with which we approach the true minimum will depend on the method of altering the list at each iteration.

Countless methods may be imagined for altering the list at each step, and one way to construct them is to imagine moving chips on a board. The squares are numbered serially for identification and then we carry out a simulated shuffling as described in Chapter 7, with an important change in the rules. Before each move is accepted, we evaluate the change of the cost function. We accept moves for which the cost function decreases. If the cost function increases, we draw a random number r and if r satisfies $r < \exp(-\delta C/T)$ we accept the move, otherwise we reject it.

Another method would be to work directly with the distribution n_q, rather than working with the squares. It would go as follows:

Prescription for Altering the List $\{n_q\}$

Step 1: Pick q at random from integers in the range 1–Q.

Step 2: Removing one chip from a square with n_q chips would decrease by 1 the number of squares with q chips and simultaneously increase by 1 the number of squares with $q-1$ chips. (We do not need to know which squares have been affected in this formulation.) These two effects can be represent by the changes

$$n_q \to n_q - 1 \quad \text{and} \quad n_{q-1} \to n_{q-1} + 1.$$

Step 3: Pick q' at random from the range 0–Q.

Step 4: Placing the chip onto a square with $n_{q'}$ chips would decrease by 1 the number of squares with q' chips and simultaneously increase by 1 the number with $q'+1$ chips. These changes can be represented by

$$n_{q'} \to n_{q'} - 1 \quad \text{and} \quad n_{q'+1} \to n_{q'+1} + 1.$$

This prescription alters the values of four members of the set $\{n_q\}$ and it retains $Q = $ constant, as the reader is to verify in Exercise 18.2. The following program fragment shows the implementation of this procedure.

```
Procedure hatch; {generates new distribution, model[1] from model[2]}
        var       i : integer;
        Begin
        Repeat
                d := randint(1, totalq);
        Until (model[2].count[d] > 0);
        model[1] := model[2]; {reset model[1] to last accepted model}
{take care of donation first}
        model[1].count[d] := model[2].count[d] - 1;
        model[1].count[d - 1] := model[2].count[d - 1] + 1;
{now find and take care of receiver}
        Repeat
                r := randint(0, totalq - 1);
        Until (model[1].count[r] > 0);
        diffcost := costchange(d, r);
        model[1].cost := model[2].cost + diffcost;
```

{evaluate cost change using original populations}
{in taking care of receiver, must not lose effect of donation}
        model[1].count[r] := model[1].count[r] - 1;
        model[1].count[r + 1] := model[1].count[r + 1] + 1;
        End; {hatch}

Exercise 18.2
Verify that $Q = \sum q n_q$ is constant if we follow the prescription in *hatch*.

After each application of such a prescription, we can compute the change of the cost function, $\delta C = C_{new} - C_{old}$, from the following exact expression:

$$\delta C = \sum \delta \ln n_q!$$
$$= \delta \ln n_q! + \delta \ln n_{q-1}! + \delta \ln n_{q'}! + \delta \ln n_{q'+1}!$$
$$= -\ln n_q + \ln(n_{q-1} + 1) - \ln n_{q'} + \ln(n_{q'+1} + 1).$$

The following program fragment shows this evaluation:

Function CostChange;
{Evaluate change of cost function implied by change of distribution}
        var   nd, ndm1, nr, nrp1 : integer;
        Begin
        nd := model[2].count[d];
        ndm1 := model[2].count[d - 1];
        nr := model[1].count[r];
        nrp1 := model[1].count[r + 1];
        costchange := ln((ndm1 + 1) * (nrP1 + 1) / (nd * nr));
        End; {CostChange}

Exercise 18.3
Verify the expression for δC, and show that it vanishes when the n_q follow a geometric series if $n_q \gg 1$. This vanishing is the condition for a minimum of C.

Having evaluated δC, we do the following:

If $\delta C < 0$: Accept the alteration and change the list accordingly.

If $\delta C > 0$: Draw a uniform random variable r $(0 < r < 1)$ and accept the alteration if $r < \exp(-\delta C/T)$. As in earlier descriptions, a large value of $\delta C/T$ implies a small probability of acceptance. This is shown in the following program fragment:

Procedure select; {Decide to either keep new model or throw it out}
{Selection is based on a probability computed from current temperature}
        var
                x, chance : real;

```
Begin
  chance := exp(-diffCost * 100 / temperature);
{The factor 100 is chosen to give convenient fraction of rejections}
    If abs(random) / 32767 < chance Then
                accept := true
                Else
                accept := false;
End; {select}
```

The parameter T is the pseudo-temperature discussed in Section 18.2. It is gradually decreased during the search. Figure 18.2 shows two histograms of n_q obtained during such a search with $N = 10$ and $Q = 10$. The search was started with all quanta in one atom, $n_{10} = 1$ and $n_0 = 9$. The upper histogram shows the status after 1 quantum had been moved and the lower shows the final, optimized result. The histogram shows the linear decrease we had found earlier from the checkerboard simulation for small values of N and Q.

Monte Carlo optimization procedures based on this type of randomized search technique are continually finding new uses in a variety of fields.

Exercise 18.4
Carry out a random search for the distribution n_q, using one of the methods described above, for the values $N = 10$ and $Q = 15$. Devise an alternative method for shuffling the quanta and show that the result is not significantly different.

The driver program for the optimization is as follows:

```
Begin {main}
    randseed := tickcount; {reset random number generator}
    finished := false;
    init; {set parameters}
```

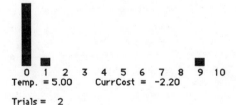

```
  0   1   2   3   4   5   6   7   8   9   10
Temp.  = 5.00       CurrCost = -2.20

Trials =   2
```

```
  0   1   2   3   4   5   6   7   8   9   10
Temp.  = 1.25       CurrCost = -7.14

Trials =  46
```

FIGURE 18.2 Search for optimum distribution of 10 quanta among 10 atoms. Upper portion shows status after first step in the search. Lower portion shows the fully optimized distribution, obtained after 46 steps in the annealing process.

```
        best := 0;  {initialize cost function}
        setTemperature;
        Repeat
            trial := trial + 1;
            ctrial := ctrial + 1;
            endCond := false;
            hatch(don, rec);  {perturb distribution [2] to new one [1]}
            select;  {decide whether to keep new model}
        If (trial MOD 20) = 0 Then
            schedule;
        If needTemp = true Then
            setTemperature
        If(accept = true) Then
            Begin
            nacceptances := nacceptances + 1;
            model[2] := model[1];
            ShowStatus;  {display current histogram}
            End;
        If model[2].cost < best Then
            Begin
            best := model[2].cost;  {update current best cost}
            model[3] := model[2];   {hold best model in model[3]}
            endFlag := 0;
            eraserect(BestRect);
            drawHistogram(3);  {draw current best histogram}
            End;
        If button Then
            endCond := true;
        Until endCond;
        ShowStatus;
        finished := true;
End.  {main}
```

REFERENCES

Binder 1984
Corey and Young 1989
Gurney 1949
Kirkpatrick et al. 1983
Metropolis et al. 1953
Whitney 1989

APPENDIXES

APPENDIX 1

THE AXIOMS OF PROBABILITY THEORY

A1.1 WHAT THE LAWS OF PROBABILITY CAN TELL US

The laws of probability do not tell us how to assign a probability to the outcome of a single event such as the tossing of a die. We must first observe each individual die and decide whether it is a "fair" one so that all sides have an equal probability of appearing. Once we have established the probability of each simple event ("spot" in the case of the die), the laws of probability tell us how to combine these probabilities into probabilities for compound events. These laws provide rules that answer the following types of question: Should we add the probabilities, multiply them, or do a combination of operations in order to predict the likelihood of a compound event?

A1.2 VENN DIAGRAM FOR COIN FLIPPING

In what follows, we shall use some terms from elementary set theory and refer the reader to Appendix 2 for more details on the concepts of sets.

The set of possible outcomes of an experiment is the *sample space,* and we can map this sample space as an area on a plane. Dividing the area into parts representing the possible outcomes, we obtain the Venn diagram for this experiment. The Venn diagram for a single flip of a coin would be an area divided into two equal parts, H and T, each part representing one of the outcomes (Figure A1.1). These two areas have two important properties:

(1) They are mutually exclusive, because they have no points in common. A coin flip cannot come up both H *and* T, only H *or* T.

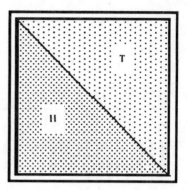

FIGURE A1.1   Venn diagram for coin flipping. Only two outcomes are possible, H and T. They fill the sample space, which is the dotted are with the double outline and will be denoted *W*.

(2) On the other hand, they are all-inclusive because they cover all the possible outcomes and their sum fills the entire sample space *W*.

The intersection of two areas, represented by ×, is the area belonging to one *and* the other. The fact that H and T have no area in common can be represented by saying that their *intersection* is null, or

$$\{H\} \times \{T\} = \varnothing, \qquad (A1.1)$$

where the symbol $\varnothing$ is the null symbol standing for zero area on the Venn diagram. Such outcomes are mutually exclusive and are said to be *disjoint*.

The *union* of two areas is the area containing all regions belonging to one *or* the other. Its symbol will be written ∪. In the experiment consisting of a single flip of a coin, H and T encompass all possible outcomes. We write this as

$$\{W\} = \{T \cup H\}. \qquad (A1.2)$$

As another example, consider the flipping of a tetrahedral (four-sided) die, illustrated in Figure A1.2. The sample space *W* is divided into four elementary events $\{T_n\}$ $(n = 1, 2, 3, 4)$, where T_n represents the event that the *n*-spot shows when the die is cast. We assume a fair die, so $P(T_n) = \frac{1}{4}$. If we now ask for the probability that the die shows a 3-spot or a 4-spot, we can apply the simple sum rule. Call this the event $C = \{T_3 \cup T_4\}$. We can compute the probabilities

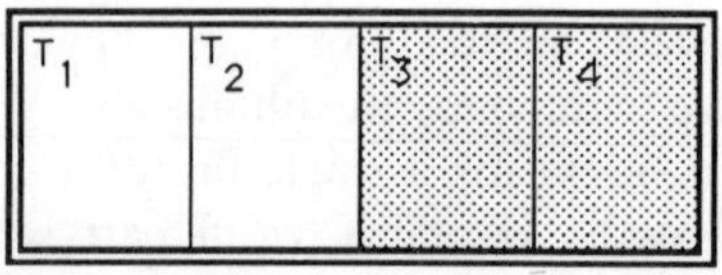

FIGURE A1.2   Elementary events T_i are disjoint and span the entire sample space. The shaded area represents the compound event $C = \{T_3, T_4\}$, and its probability satisfies $P(C) = P(T_3 \cup T_4)$.

for this event from the sum rule for disjoint events. In this case, we have

$$P(T_3) = P(T_4) = \tfrac{1}{4}, \qquad P(C) = P(T_3 \cup T_4) = \tfrac{1}{4} + \tfrac{1}{4} = \tfrac{1}{2}.$$

A1.3   MONTE CARLO EVALUATION OF AREAS ON A VENN DIAGRAM

There is a useful analogy between probabilities and areas on the Venn diagram. Suppose we were to throw darts at a diagram like Figure A1.2 in such a way that all darts hit the target, but they hit it in a randomly chosen spot. In this case, we would expect about half of the darts to hit C and half to miss it. This illustrates the relationship between area and probability in such a diagram.

Figure A1.3a shows a Venn diagram with a more interesting shape—a square with an inscribed circle. The square is 2 units long on each side. For the purpose of computation, we consider one-quarter of this diagram as a target (Figure A1.3b) and simulate throwing darts at it with a computer. In each "throw" the computer chooses two random numbers between 0 and 1.0, corresponding to the distances rightward from the vertical y-axis and upward from the x-axis. If the dart falls inside the circle, we call it a "hit," and we count the proportion of hits. In a long series of throws we expect the number of hits to be proportional to the area of the circle, so

$$\frac{\text{Hits}}{\text{Hits} + \text{misses}} = \frac{\text{Area of circle}}{\text{Area of square}} = \frac{\pi}{4}.$$

This simulation could be used to derive a Monte Carlo evaluation of π, and

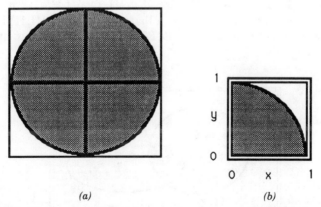

(a)   *(b)*

FIGURE A1.3   Monte Carlo evaluation of area. The area of the circle is evaluated by locating points at x and y chosen at random in the range 0–1. The fraction of points lying inside the circle ($x^2 + y^2 < 1$) determines the relative area.

a similar process can be used to evaluate an area of *any* shape, as long as the computer can decide whether a hit has occurred.

A1.4 EVENTS THAT ARE NOT DISJOINT

Before describing the laws of probability formally, let us use our intuition for a slightly more complicated case. Suppose two of the possible outcomes (A and B) of an event overlap, as in Figure A1.4. The area of overlap is indicated by $A \times B$; it is the intersection. Note that these two events do not cover the entire sample space, so they do not cover all possible outcomes. The important aspect of this diagram is that the area of the union, $C = A \cup B$, is less than the sum of the areas of A and B. In fact, we can see that the following holds:

$$\text{Area}(C) = \text{Area}(A) + \text{Area}(B) - \text{Area}(A \times B). \tag{A1.3}$$

So the area is the sum reduced by the intersection, because the area of the intersection is counted twice if we simply sum the areas of A and B.

To see how this applies to probabilities, let us suppose a coin is flipped twice. The possible outcomes are $\{H, H\}$, $\{H, T\}$, $\{T, H\}$ and $\{T, T\}$. The sample space for this experiment is shown in Figure A1.5, where the rows (horizontal) are labeled with the outcome of the second toss and the columns (vertical) are labeled with the outcomes of the first toss.

Now let us define the event C to correspond to tossing H at least once in the two tosses. To compute $P(C)$, our first guess might simply be to apply the sum rule and add the probabilities of H on the first toss and H on the second toss. But this would give $P(C) = \frac{1}{2} + \frac{1}{2} = 1$, which cannot be correct, because the outcome C cannot have unit probability, corresponding to certainty. The error was to add the probabilities of events that are not mutually exclusive. The effect was to include the area 4 twice. We added $P(\{H, T\}, \{H, H\}) = \frac{1}{2}$ to $P(\{T, H\}, \{H, H\}) = \frac{1}{2}$, obtaining unity. But doing it that way, the event $\{H, H\}$ is included twice, so we should have subtracted $P(\{H, H\}) = \frac{1}{4}$, obtaining the correct answer $\frac{3}{4}$.

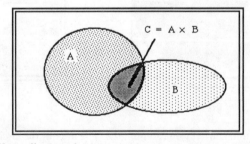

FIGURE A1.4 Venn diagram for two overlapping outcomes. The area of overlap is the intersection $C = A \times B$.

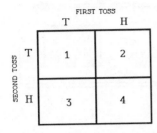

FIGURE A1.5 Venn diagram for the toss of two coins. The possible outcomes of the first toss are the column labels and the outcomes of the second toss are the row labels. For example, area 4 corresponds to {H, H} and area 3 corresponds to {T, H}.

To avoid this type of pitfall, we formulate the sum rule so that it applies to overlapping as well as mutually exclusive outcomes.

A1.5 THE FUNDAMENTAL AXIOMS

Four terms that we have been using require careful definition:

A *sample space* is the set of all possible outcomes of an experiment.

An *experiment* is a particular type of probabilistic event, such as the flipping of a coin or the drawing of a card. Different experiments have different sample spaces.

A *trial* is single execution of the experiment. A trial would be an actual toss of a coin.

An *outcome* is the result of a trial; it consists of specifying the region of sample space that was selected by the trial. For example, H might be the outcome of a single coin toss.

We now state the axioms of probability:

Axiom I: *To each possible outcome of an experiment we can assign a non negative real number P(A), which will be called the "probability of the event A."*

Axiom II: *The probability associated with the entire sample space W is unity:* $P(W) = 1.$ We refer to W as the "certain" event. In tossing a coin, there are just two possible outcomes, H and T, so their union is certain to occur, $P(H \cup T) = P(W) = 1.$

Axiom III: *If A and B are disjoint, the probability of their union is the sum of their probabilities.*

That is, if the intersection of A and B is the null set, $A \times B = \emptyset$, then

$$P(A \cup B) = P(A) + P(B). \tag{A1.4}$$

This extends at once to any number of disjoint events.

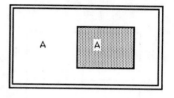

FIGURE A1.6 The complement of A is A', and it encompasses the portion of the sample space that is not in A. The probabiliity of A is unity minus the probability of its complement.

Corollary: *We can now write down the sum rule for events that intersect, that is, that have an overlapping area, $A \times B$, in the Venn diagram of Figure A1.4. The area of the union $A \cup B$ is just the sum of the areas minus the overlap. We can write this as*

$$P(A \cup B) = P(A) + P(B) - P(A \times B). \tag{A1.5}$$

Corollary: *Let A' be the* complement *of A with respect to the sample space W. The complement is the area that is not in A (Figure A1.6). Then the union of A and its complement A' fill the sample space, so we have*

$$P(W) = P(A) + P(A') = 1$$

and

$$P(A') = 1 - P(A). \tag{A1.6}$$

Sometimes it is simpler to compute the probability that an event will *not occur* and then subtract this probability from unity. As an example, let us compute the probabiliity that in a group of five people, at least two of them will have their birthday on the same day of the year. Call this event A and its probability $P(A)$. The most direct way to compute $P(A)$ is to consider the complementary problem and compute $P(A') = 1 - P(A)$.

What is the probability that no two people in a group of five will share a birthday?

To compute this, consider the people one at a time. Whatever the birthdate of the first happens to be, the probability that the second will not share that birthday is $\frac{364}{365}$. The probability that the third will not share either of the first two birthdays is $\frac{363}{365}$. Likewise the probabilities that the fourth and fifth will not share birthdays with the others are $\frac{262}{365}$ and $\frac{361}{365}$. The probability that none will share birthdays is the product of these four factors, $P(A') = 0.973$. Therefore the probability that at least two will share birthdays is $P(A) = 0.027$.

Exercise A1.1
How many people must be in the room in order for the probability of shared birthdays to exceed 0.5? Most people find the result quite surprising.

A1.6   CONDITIONAL PROBABILITY

Table A1.1 is a list of the 30 brightest stars in the sky, with their magnitudes and coordinates on the celestial sphere. (Magnitude is a measure of apparent brightness and is such that brighter stars have smaller magnitudes. Right ascension and declination are equivalent to longitude and latitude. Negative declination indicates the southern hemisphere.)

We divide the stars into two classes according to whether they are brighter or fainter than magnitude 1.0. Of the 30 stars, 14 have magnitudes less than 1.0; call these the brighter stars (class B). The remaining are the fainter stars (class F). If $P(B)$ is the probability that a randomly selected star is of class B,

TABLE A1.1   The 30 Brightest Stars

Name	Magnitude	Right Ascension		Declination	
1. Sirius	−1.47	6 h	43 m	−16°	39′
2. Canopus	−0.73	6	21	−52	40
3. Vega	0.04	18	35	+38	44
4. Arcturus	0.06	14	13	+19	27
5. α Centauri	0.06	14	36	−60	38
6. Rigel	0.08	5	12	−8	15
7. Capella	0.09	5	13	+45	57
8. Procyon	0.34	7	37	+5	21
9. Achernar	0.47	1	36	−57	29
10. Hadar	0.59	14	00	−60	08
11. Altair	0.77	19	48	+8	44
12. Betelgeuse	0.80	5	52	+7	24
13. Aldebaran	0.86	4	33	+16	25
14. Spica	0.96	13	23	−10	54
15. α Crucis	1.05	12	24	−62	49
16. Antares	1.08	16	26	−26	19
17. Pollux	1.15	7	42	+28	09
18. Fomalhaut	1.16	22	55	−29	53
19. β Crucis	1.24	12	45	−59	25
20. Deneb	1.26	20	40	+45	06
21. Regulus	1.35	10	06	+12	13
22. Adhara	1.50	6	57	−28	54
23. Castor	1.58	7	31	+32	00
24. γ Crucis	1.62	12	28	−56	50
25. Shaula	1.62	17	30	−37	04
26. Bellatrix	1.64	5	22	+6	18
27. El Nath	1.65	5	23	+28	34
28. Miaplacidus	1.67	9	13	−69	31
29. Alnilam	1.70	5	34	−1	14
30. Avior	1.85	8	21	−59	21

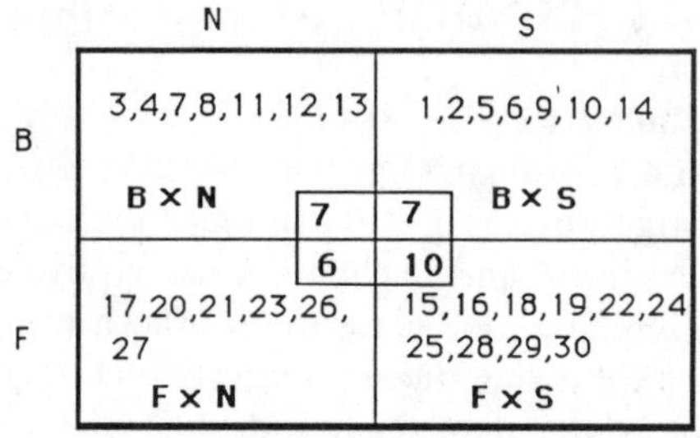

N   S

	3,4,7,8,11,12,13	1,2,5,6,9,10,14
B	B × N 7	7 B × S
	6	10
F	17,20,21,23,26, 27	15,16,18,19,22,24 25,28,29,30
	F × N	F × S

FIGURE A1.7 The two-way classification of the brighter stars according to brightness class (B, F) and hemisphere (N, S). The number of stars in each class is indicated in the central box and the intersections defining each class are also shown.

and $P(F)$ is the probability of belonging to class F, then as all stars in the list must belong to one or the other we have $P(B) + P(F) = 1.0$, and $P(B) = \frac{14}{30}$, $P(F) = \frac{16}{30}$.

The stars may also be divided into two classes (N and S) according to whether they lie in the northern or southern hemisphere. This two-way classification is represented in Figure A1.7, which assigns each star number to one of the four distinct classes.

By counting the fraction of stars in each class we can assign the following probabilities that a star randomly chosen from the list will belong to that class.

$$P(N) = \tfrac{13}{30},$$
$$P(S) = \tfrac{17}{30},$$
$$P(B) = \tfrac{14}{30},$$
$$P(F) = \tfrac{16}{30}.$$

Exercise A1.2
Scan the list in Table A1.1 and estimate the probabilities of belonging to each subset, then count them and show that the above probabilities hold.

For the regions of sample space represented by various intersections we have

$$P(F \times N) = \tfrac{6}{30},$$
$$P(F \times S) = \tfrac{10}{30},$$
$$P(B \times N) = \tfrac{7}{30},$$
$$P(B \times S) = \tfrac{7}{30}.$$

From these we may compute, for example, the probability of being in the north (N) or belonging to the fainter (F) group,

$$P(F \text{ or } N) \equiv P(F \cup N) = P(F) + P(N) - P(F \times N) \qquad (A1.7)$$

$$= \frac{16 + 13 - 6}{30} = \frac{23}{30}.$$

With these star data we can consider another type of problem: Given that a star is in the north, what is the probability that it will be in the brighter group? This is called the *conditional probability* because the outcome of the trial assumes that a previous condition has been met; namely, the star is to be selected from the northern hemisphere. We write this probability with the symbol $P(B|N)$, and by examining the figure, we can see that the value is

$$P(B|N) = \tfrac{7}{13}.$$

We now formalize the process of evaluating conditional probabilities.

Definition: The conditional probability of B under the condition N is the probability that B will occur when N is known to have occurred. It is written $P(B|N)$. The given *conditioning* event N appears second in the parentheses, and we refer to B as the *conditioned* event. An example of the notation is

$$P(\text{star is bright } given \text{ it is in north}) = P(B|N).$$

With the aid of Figure A1.7, we can see that the conditional probability of a northern star being in the brighter subset is just the probability of its being in the north *and* in the brighter subset divided by the probability of its being in the north. Thus the following relationship holds:

$$P(B|N) = \frac{P(N \times B)}{P(N)} = \frac{7/30}{13/30} = \frac{7}{13}.$$

Thus the conditional probabilities are derived from the probability of the joint event divided by the probability of the conditioning event. In this example, the conditioning event was the selection of the northern half of the sky.

Conditional probabilities obey the arithmetic of ordinary, unconditional probabilities as long as they refer to the same conditioning event; that is, as long as A is fixed in expressions like the following:

$$P(B|A) + P(F|A) = 1. \tag{A1.8}$$

A1.7   CAUTION ON INTERPRETING CONDITIONAL PROBABILITIES

A large numerical value for a conditional probability implies that the two events are correlated; that is, they appear together—if one occurs, the other can be expected. Sometimes this correlation reveals that one event causes the other, the way a cloudy sky might cause rainy weather. Quite often, however, events occur together because they are the consequences of other events, not of one another. (The wearing of raincoats and the growth of crops may correlate with the appearance of cloudy skies, but no one would claim they cause each other.) In short, correlation is a necessary but not a sufficient condition for inferring the existence of a causal connection.

A1.8   THE MULTIPLICATION RULE

From the relationship defining conditional probability, we find the following very useful "multiplication rule" giving the probability of the simultaneous occurrence of A and B,

$$P(A \times B) = P(B|A)P(A). \qquad (A1.9)$$

The meaning of the multiplication rule may be stated in words as follows:

The probability of A and B occurring together is the probability of B given A times the probability of A.

This is also called the *joint probability*. As an example, the joint probability that an arbitrarily chosen star among the 30 brightest is in the northern hemisphere *and* is brighter than magnitude 1.0 may be expressed as $P(B \times N)$. We may evaluate this with the multiplication rule and the probabilities in Exercise A1.2. We have

$$P(B \times N) = P(B|N)P(N) = (7/13)(13/30) = 7/30,$$

as may be verified by counting the number of stars in the intersection of A and B.

A1.9   STATISTICAL INDEPENDENCE

If the occurrence of event A is uncorrelated with another event B, it is said to be *statistically independent* of it. The relationship may be expressed by the equality

$$P(A|B) = P(A), \qquad (A1.10)$$

which says that the probability of A *conditioned by B* is the same as the *total* probability of A.

If we insert this into the multiplication rule,

$$P(A \times B) = P(B|A)P(A),$$

we find

$$P(A \times B) = P(B)P(A). \qquad (A1.11)$$

Thus, if two events are statistically independent, the probability of their joint occurrence is just the product of their individual probabilities. This is the rule we used in Chapter 1, without justification, when we computed the probability of two heads in a row, for example. We assumed that each toss was statistically independent of the previous ones. This relationship is also useful as a test for *statistical independence*. If it is found to be true for two events, they are statistically independent.

A1.10   THE ASSIGNMENT OF PROBABILITIES

The axioms and corollaries tell us nothing about the actual assignment of probabilities to specific events. They merely describe the rules by which the numerical values of the probabilities may be manipulated, once they have been assigned. To assign probabilities, we must go into the physical world and examine the process we are modeling. We can perform experiments and derive the probabilities empirically, or we can evaluate them from a theory.

Here is an example of the empirical method of assigning probabilities. In his classic description of his single-handed voyage around the world, Joshua Slocum describes spending a few nights in the Straits of Magellan a short distance from shore. In order to avoid being taken unawares by natives, who were infamous for sneaking aboard vessels at anchor while their crews slept, he scattered thumbtacks on the deck before going to sleep. He was not disappointed. During the night, he heard loud cries of pained anger from several barefoot natives as they leapt overboard to escape the tacks.

Suppose we set out to compute the fraction of tacks, f, that had landed point-upward on the deck and the fraction, $f' = 1 - n$, that lay with their points against the deck. We are immediately stumped, because the theoretical calculation seems hopelessly complicated. To carry it out convincingly, we would have to include details of spin as the tacks fell, the nature of the bounce, the resilience of the deck, and so on. Clearly an empirical approach is needed, so we take a fistful of tacks and drop them. After counting them, we drop the fistful again. Gradually a fraction appears to emerge and we finally get to the point where we say, "That is close enough." But how can we know that Slocum's tacks looked anything like the sample we tested, or that his deck was similar to ours? We cannot. We can only say, "If his tacks and deck had been similar to ours…"

So the empirical method will require a leap of faith if we are to apply it to a probability calculation.

APPENDIX 2

PROPERTIES OF DISTRIBUTIONS

A2.1 LISTS, SETS, AND POPULATIONS

In this appendix, we shall systematize some of the terminology to describe the relative frequencies attached to the domains of a sample space. We start with a small amount of terminology.

List A list is an ordered group of symbols. It can contain repetitions of a given element. When we write down the outcomes of a coin-flipping experiment in the order in which they occur, we create a list.

Set A set is a single entity comprising a collection of items. A set may be described by naming all its elements or by describing its elements in terms of their common property. The members of a set are not ordered. The seven brightest stars in the Pleiades constitute a set. (See Figure A2.1.) Any arbitrarily chosen star in the sky either *is* or *is not* a member of that set.

Population Now suppose that we carry out an experiment and count the relative frequencies for each outcome. The resulting set of events and their associated frequencies will be called a *population*. A population can be considered as the summary of a list of outcomes in which each outcome is tagged with a frequency.

We must distinguish between a *sample* population and a *parent* population. A parent population is the expected result of performing an experiment an unlimited number of times. It is an abstraction beyond experience. A sample population describes the results of a finite number of trials. The frequencies

289

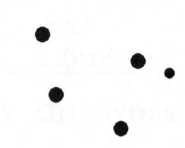

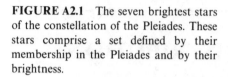

FIGURE A2.1 The seven brightest stars of the constellation of the Pleiades. These stars comprise a set defined by their membership in the Pleiades and by their brightness.

TABLE A2.1   Summary of Flipping a Coin

List of outcomes	H H T H H T T T H T T H H H H T T T T T		
Set of possible outcomes	H T		
Sample population	Outcomes (frequencies):	H(9/20)	T(11/20)
Parent population	Outcomes (frequencies):	H(10/20)	T(10/20)

assigned to each outcome is a sample population will vary from one sample to another if the experiment has an element of chance. This is not the case for the parent population. In a parent population, frequencies are uniquely assigned on the basis of the expected outcome of an unlimited number of trials—usually derived from a theory for the expected behavior.

These distinctions are illustrated in Table A2.1, describing a coin-flipping experiment. The frequencies for the parent population in this case are derived from the assumption that the coin is equally likely to land on either of its two faces.

The quantitative description of a population is the probability distribution, and this will be the focus of the current chapter. The first paragraph of the next section will provide an example.

A2.2   DISCRETE DISTRIBUTIONS

(a) Density Distributions

This paragraph is a list of words, and each word is recognized by the list of letters it contains. Suppose we were to focus attention, not on the meaning of the paragraph, but on its more concrete statistical properties. We could attach a specific attribute to each word, such as its length, and we might count the words of various lengths. In this way, we might construct a sample population from the paragraph. Instead of considering its significance as a whole, we would consider it as a set of words, and we would characterize this set by the frequencies of different word-lengths.

TABLE A2.2   Density Distribution of Word-Lengths

Length, L	1	2	3	4	5	6	7	8	9	10	11	12
Number, $n(L)$	5	26	17	14	11	3	7	4	7	3	3	2

Table A2.2 and Figure A2.2 display the numbers of words of different lengths in the preceding paragraph. To each word-length L we have assigned a number $N(L)$ equal to the number of times that length occurred in the paragraph. These are the numbers plotted in the figure, and the resulting plot shows the distribution of words of different lengths. This is also known as the density distribution of word-lengths. It is a discrete distribution because each length L is an integer and can take only a discrete set of values. For this reason, it is plotted as a bar graph (histogram) rather than a continuous curve.

We shall speak of the distribution of word-lengths as a statistical quantity because it refers to the frequency of a particular attribute, L. It is unaffected by the order of the words, so it is a characteristic of the paragraph as a population rather than as a list. We can make a statistical comparison of word-lengths in different paragraphs, or in different books, with no regard for the placement of the paragraph in the book or the nature of the book. These distributions can be the subject of probability statements.

(b) Cumulative Distributions

Often we are interested in the number of items in a population that exceed a certain condition, rather than matching it. For example, we might want the number of words in a paragraph that are equal to or shorter than length L.

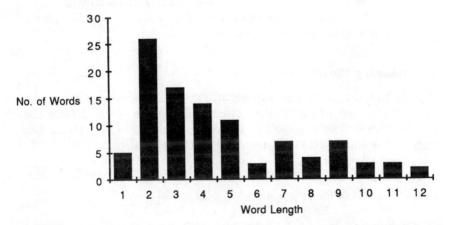

FIGURE A2.2   Histogram of word-lengths in the first paragraph of this section, based on the data in Table A2.2.

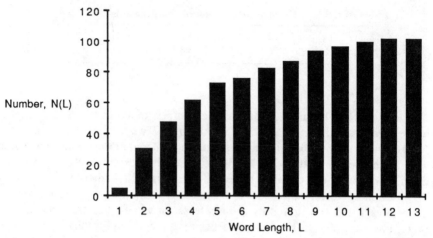

FIGURE A2.3   Cumulative distribution $N(L)$ of words shorter than or equal to L.

This is the cumulative distribution $N(L)$, and we may derive it from the density distribution $n(L)$, from the summation,

$$N(L) = \sum_{j=1}^{L} n(j). \qquad (A2.1)$$

For example, from Table A2.2, we easily find $N(3) = 48$, so there are 48 words of length 1–3 letters.

The $n(L)$ are all positive, so $N(L)$ must increase monotonically with increasing L. When the largest value of L is achieved, all subsequent values of $n(L)$ vanish, so $N(L)$ becomes flat. This general behavior is true of all cumulative distributions; they have no secondary peaks and valleys. Figure A2.3 shows the cumulative distribution $N(L)$ derived from Table A2.2.

(c) Probability Distributions

A probability distribution gives the relative frequencies with which the associated event is expected to occur. Such distributions can be derived from density distributions, and we now state the two conditions that must be met if a discrete density distribution $p(L)$ is to qualify as a probability distribution:

(1) $p(L)$ is in the range 0 to 1 for all L: $0.0 \leqslant p(L) \leqslant 1.0$.
(2) The sum of the $p(L)$ is unity: $\sum p(L) = 1.0$.

Returning to the word-length data in Table A2.2, we may generate a distribution $p(L)$ satisfying these conditions from the density distribution $n(L)$

if we *normalize* them. That is, we divide each $n(L)$ by N, the total number of words in the paragraph. This normalization guarantees that the probability distribution will sum to unity and satisfy the second condition. Thus, if

$$N = \sum_L n(L),$$

we may define each quotient

$$p(L) \equiv \frac{n(L)}{N} \qquad\qquad (A2.2)$$

as the *frequency* of the corresponding word-length. The shape of the distributions $P(L)$ and $n(L)$ are obviously the same.

A2.3   CHARACTERIZING A DISTRIBUTION NUMERICALLY

In addition to plotting a distribution, we often wish to characterize its quantitative properties in terms of a few numbers. We now consider various ways of doing this, and as an example we take the work-length data in Table A2.2.

(a) Mean

The mean is an average that takes into account the number of words of different lengths. It is a "weighted" sum in which the weighting is proportional to the frequency of each length. We shall use the symbol $\langle L \rangle$ to indicate the mean of a sample distribution, and it is derived from

$$\langle L \rangle = \frac{1}{N} \sum L n(L) = \sum L p(L). \qquad\qquad (A2.3)$$

With the data in Table A2.2, we easily verify that the mean word-length is $\langle L \rangle = 4.6$.

(b) Mode

The mode is the value of the argument L, corresponding to the greatest number of words. It corresponds to the highest point of the distribution $n(L)$. In this case $\text{Mode}(L) = 2$.

(c) Median

The median word-length is the value of L that lies halfway between the smallest

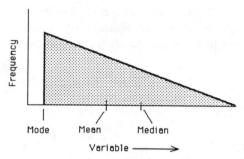

FIGURE A2.4   Three characteristics of a distribution, as defined in the next. For the triangular distribution, the mode, mean, and median are distinct.

and the largest values of L. It is given by

$$\text{Median}(L) = \tfrac{1}{2}[\text{Max}(L) + \text{Min}(L)].\qquad(A2.4)$$

In this case, the median word-length is 6.5.

The differences between the mode, median, and mean depends on the shape of the distribution, as illustrated in Figure A2.4. If the distribution is symmetric, these three quantities will be equal. When it is not symmetric, the median will usually depart farther from the mode than the mean does. None of these numbers gives a useful measure of the width of the distribution. Even the spread may be misleading if the distribution has two uncharacteristic "outliers." We now look for other ways to characterize the effective width of $n(L)$.

(d) Variance

When we set out to analyze the width of a distribution, many possibilities are open to us. The most direct would be to find a measure of the average distance of the points from the mean value. How can we best do this? Of course, some of the deviations from the mean will be positive and others negative, and their average is zero by definition of the mean. So the average of the deviations is not a useful measure of width. Another possibility would be to take the average of the absolute value of the deviation, that is, the average without regard to sign. This is easy to calculate for a single distribution, but it is not convenient when we deal with sums of distributions.

The following procedure leads to the "variance," which is not only a convenient measure to compute but, as we shall see later, is often a meaningful measure of the spread.

To evaluate the variance we first compute the mean $\langle L \rangle$. Next, we subtract $\langle L \rangle$ from each L, square each difference, and then calculate the mean of these squared differences. This mean is calculated by summing over the frequencies of occurrence. We designate the variance as s^2, and we have

$$s^2 = \frac{1}{N}\sum (L - \langle L \rangle)^2 n(L),$$ (A2.5)

where the total number is

$$N = \sum_L n(L).$$

Exercise A2.1
From the values of L and $n(L)$ listed in Table 2.2, find the mean word-length and show that the variance is $s^2 = 8.20$.

To derive a relationship that will make the task of computing the variance slightly easier, we start from the definition of the variance and expand the binomial under the summation. Using $p(L) = N(L)/N$, we have

$$s^2 = \sum (L - \langle L \rangle)^2 p(L)$$

and we find

$$s^2 = \sum (L^2 - 2L\langle L \rangle + \langle L \rangle^2) p(L)$$
$$= \sum L^2 p(L) - 2\langle L \rangle \sum L p(L) + \langle L \rangle^2 \sum p(L).$$

The first term is the mean of the square, $\langle L^2 \rangle$. And because $\sum p(L) = 1$, the third term is the square of the mean. It combines with the second term to give

$$s^2 = \langle L^2 \rangle - \langle L \rangle^2.$$ (A2.6)

This important relationship says that the variance is the difference between the mean of the square and the square of the mean for any variable. No assumptions have been made about the shape of the frequency distribution $p(L)$, so this relationship will apply to any variable.

(e) Standard Deviation

The square root of the variance, $\sqrt{s^2}$, is also known as the root-mean-square deviation from the mean or the standard deviation. It is indicated by the greek letter sigma, σ. Thus

$$\sigma \equiv \sqrt{s^2}$$ (A2.7)

is a measure of the width of the bulk of a distribution.

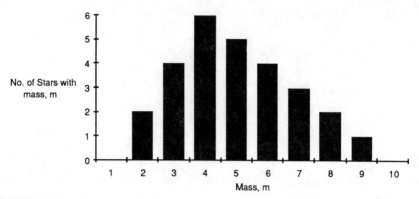

FIGURE A2.5 Triangular distribution function for illustrating computation of mean, mode, median, and variance in Exercise A2.2.

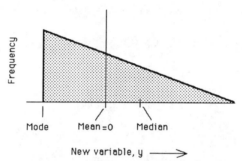

FIGURE A2.6 Illustration for Exercise A2.3. Similar to the distribution in Figure A2.5, except the abscissa is the transformed variable $y = x - \langle x \rangle$, so $\langle y \rangle = 0$. The variance is unchanged.

Exercise A2.2

A triangular distribution is shown in Figure A2.5. The variable x represents the numerical value of a property such as mass, and $n(x)$ is the number of stars with that value.

(a) Compute the corresponding probability density distribution $p(x)$, using the definition given earlier in this chapter.

(b) Evaluate the mean, median, and mode of $p(x)$.

(c) Evaluate the variance of $p(x)$ using the definition as well as the simplified formula.

Exercise A2.3

Starting with the definition $s^2 = \langle x^2 \rangle - \langle x \rangle^2$ and the transformation $y = x - \langle x \rangle$, illustrated in Figure A2.6, show that

$$s^2 = \langle y^2 \rangle - \langle y \rangle^2.$$

Thus the variance of y (its mean square value) is the same as the variance of the original variable x.

A2.4   FITTING A CONTINUOUS PROBABILITY FUNCTION TO A HISTOGRAM

Experimental data are usually divided into bins, producing histograms. Theories, on the other hand, often produce continuous curves. One way to compare a theory with a set of data is to fit the curve to the histogram. As an example, we consider the equation describing a continuous density distribution that often arises in the analysis of errors,

$$n(x) = A \exp\left(-\frac{(x-m)^2}{2s^2}\right).\tag{A2.8}$$

This is the bell-shaped "normal" distribution function, illustrated in Figure A2.7 and discussed in Chapter 12. It is a symmetric curve whose maximum occurs at $x = m$; its width is characterized by the spread $\sigma = \sqrt{s^2}$, which we have defined as the square root of its variance.

The quantity $n(x)\,dx$ is often used to represent the number of measurements that fall in the interval $(x, x + dx)$. Where the curve is high, a larger number of measurements are expected to fall. We may convert the function to a probability density $p(x)$ by normalizing, that is, by choosing the coefficient A so that

$$\int_{-\infty}^{\infty} n(x)\,dx = 1,$$

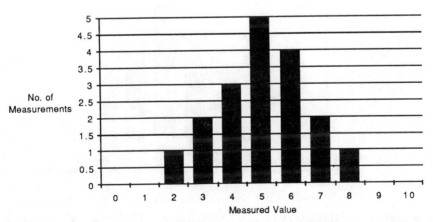

FIGURE A2.7a Illustrative histogram of error data to be fitted with a normal distribution. Each point indicates the number of data points with each measured range, $x - 0.5$ to $x + 0.5$.

or

$$A \int_{-\infty}^{\infty} \exp\left(-\frac{(x-m)^2}{2s^2} \right) dx = 1,$$

which gives

$$A = \frac{1}{\sqrt{2\pi s^2}},$$

so the probability that x occurs in the interval $(x, x + dx)$ is written

$$p(x)\, dx = \frac{1}{\sqrt{2\pi s^2}} \exp\left(-\frac{(x-m)^2}{2s^2} \right) dx. \tag{A2.9}$$

Now suppose we have a set of data, n_i, represented by the histogram in Figure A2.7a. These might represent the individual measurements of the brightness of a star, divided into bins, $x - 0.5$ to $x + 0.5$. Our task is to find the best-fitting normal curve through these points, so that we may see how well they conform to the expected distribution. This fitting requires finding from the data the best values of m and s, because these two parameters completely define the normal curve.

We start by normalizing the histogram so it represents a probability density, setting $p_i = n_i = N$. This merely alters the y-scale of the histogram, with the result shown in Figure A2.7b.

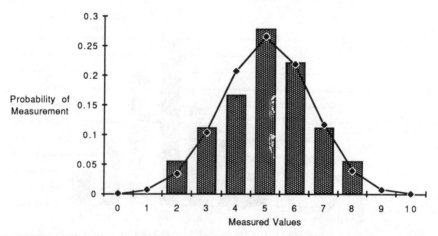

FIGURE A2.7b   Histogram of Figure A2.7a but renormalized to produce a probability density function. The normal curve derived in the text is shown superposed on the histogram to provide a comparison between the data and the shape of the theoretical distribution.

Next we identify m in the expression for $p(x)$ with the mean of the data. That is, $m = \sum i p_i$. Carrying out the summation with the data of Figure A2.7a, we find for the mean $m = 5.06$. Finally, we fit the width of the histogram by equating its variance to the variance of the normal distribution. That is, we set up the equality

$$\sum (i - m)^2 p_i = \int_{-\infty}^{\infty} (x - m)^2 p(x)\, dx.$$

Inserting the expression for $p(x)$, we find that the integral equals s^2, so

$$\sum (i - m)^2 p_i = s^2. \tag{A2.10}$$

Carrying out the summation with the data in Figure A2.7a, we find

$$s^2 = 2.27 \quad \text{and} \quad \sigma = 1.51.$$

Having thus determined m and s^2 from the histogram, we can draw the normal curve, and the result is shown in Figure A2.7b.

A2.5   DRAWING NUMBERS RANDOMLY FROM A DISTRIBUTION

Occasionally we need to draw a random variable from a population obeying a given probability law. For example, we may wish to simulate errors whose sizes obey the normal distribution or the lifetimes of individual particles that obey an exponential distribution.

We start from the cumulative distribution $P(x)$, and if it is not available, we construct it from the probability density $p(x)$, using the methods described earlier in this chapter. Remember, that $P(x)$ varies monotonically from 0 to 1 over the range of its random variable x. The relationship between x and $P(x)$ for a normal distribution with a mean $m = 2$, variance $s^2 = 0.5$, and spread $\sigma = 0.707$ is shown in Figure A2.8.

In the rare case that the expression for the cumulative distribution $P(x_i) = y_i$ can be solved analytically, giving $x_i = G(y_i)$, we simply draw values of $0 < y_i < 1$ from a uniform random distribution and compute each corresponding x_i. These x_i are distributed with the appropriate cumulative distribution. For example, if we are drawing from an exponential density distribution $p(x) = e^{-ax}$, we first find the cumulative distribution $P(x) = 1 - e^{-ax}/a = y_i$. Then we solve for x_i, finding

$$x_i = -\frac{1}{a}[\log_e(a) + \log_e(1 - y_i)].$$

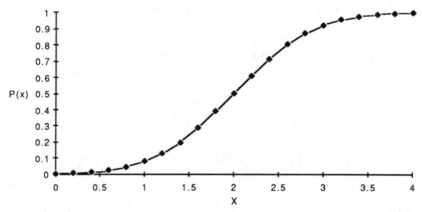

FIGURE A2.8   Cumulative probability for the normal distribution with mean of 2.0 and variance of 0.5 is plotted along the y-axis. This is also known as the probability integral. The steeper portion of the cumulative distribution $P(x)$ corresponds to the more frequent values, as those values of x would appear more often if $P(x)$ is drawn from a uniform random distribution. The slope of $P(x)$ is the probability density, which in this case is the familiar bell-shaped curve.

Drawing y_i from a uniform random distribution, we compute each x_i from this expression. The resulting values of x_i will follow an exponential density distribution.

More typically, the equation cannot be solved, and we must resort to setting up a table to give the transformation $y_i \rightarrow x_i$. This can be achieved by dividing the range of x_i into convenient bins and computing the bin boundaries from the corresponding values of $P(x_i) = y_i$. Then we draw values of y_i in the range $0 < y_i < 1$ from a uniform random distribution, find the corresponding bin, and read the corresponding x_i from the table.

A2.6   CHEBYSHEV INEQUALITY AND LAW OF LARGE NUMBERS

Two useful results concerning distributions may be obtained by elementary methods. These results apply only to distributions that have a finite second moment, but they do not depend on the detailed shape of the distributions. They both indicate how the bulk of a distribution is confined by its variance.

Suppose there is a random variable X that takes the values a_i with probability p_i; that is,

$$P(X = a_i) = p_i \qquad (i = 1, 2, 3, \ldots).$$

Then the mean is

$$\langle X \rangle = \sum_{i=1}^{\infty} a_i p_i, \qquad (A2.11)$$

TABLE A2.3   Spreadsheet Calculation of Partial Sums Described by the Chebyshev Inequality[a]

	A	B	C	D	E	F
1	i		a	p	a*a	p*a*a
2	1		-3.00	0.01	9.00	0.09
3	2		-2.00	0.03	4.00	0.12
4	3		-1.00	0.10	1.00	0.10
5	4		-0.50	0.17	0.25	0.04
6	5		0.00	0.38	0.00	0.00
7	6		0.50	0.17	0.25	0.04
8	7		1.00	0.10	1.00	0.10
9	8		2.00	0.03	4.00	0.12
10	9		3.00	0.01	9.00	0.09
11				1.00	(x*x)=	0.71
12						
13						
14	E, epsilon	0	0.5	1	2	3
15	Partial Sum	0.71	0.62	0.42	0.18	0.00
16	(x*x)/(E*E)		2.82	0.71	0.18	0.00
17	P(|x|>E)		0.28	0.08	0.02	0.00

[a]Several representative values of epsilon (ε) are listed.

and the second moment is

$$\langle X^2 \rangle = \sum_{i=1}^{\infty} a_i^2 p_i. \tag{A2.12}$$

We assume it to be finite. Table A2.3 shows an example, computed with a spreadsheet program.

It is clear that a sum of positive numbers over a restricted sample is no greater than the sum over the entire range. hence, if we choose any $\varepsilon > 0$, we have

$$\sum_{i=1}^{\infty} a_i^2 p_i \geqslant \sum_{|a_i| > \varepsilon} a_i^2 p_i. \tag{A2.13}$$

These partial-range sums are given in Table A2.3 and should be verified by the reader. Furthermore, we have the inequality

$$\sum_{|a_i| > \varepsilon} a_i^2 p_i \geqslant \varepsilon^2 \sum_{|a_i| > \varepsilon} p_i = \varepsilon^2 P(|X| > \varepsilon). \tag{A2.14}$$

By chaining these inequalities, we find the Chebyshev inequality:

$$\langle X^2 \rangle > \varepsilon^2 P(|X| > \varepsilon)$$

or

$$P(|X| > \varepsilon) < \frac{\langle X^2 \rangle}{\varepsilon^2}. \tag{A2.15}$$

In words, this states that, for any positive ε, the probability that the variable lies outside ε is less than the ratio of the variance to ε^2. This gives a rough-and-ready estimate of the probability of finding certain values of X, with a minimum of knowledge about the distribution itself. Table A2.2 verifies this inequality and illustrates its quantitative meaning for a particular distribution. (In this case, the inequality is not a very strong one.)

From this we may derive the law of large numbers, which states that the mean of a sample population tends toward the expectation value of the parent population as the sample size increases. We start by defining S_n as the sum of n variables drawn from a parent population whose mean is m. Again we chose an arbitrary small number ε, then the law of large numbers can be expressed as

$$P\left(\left|\frac{S_n}{n} - m\right| \geqslant \varepsilon\right) \to 0, \quad \text{in the limit as } n \to \infty. \tag{A2.16}$$

In words, the difference between the sample mean S_n/n and the parent mean m has a vanishing probability of exceeding ε as the sample size increases without limit. To prove this, we rewrite the expression as

$$P\left(\left|\frac{S_n}{n} - m\right| \geqslant \varepsilon\right) = P\left(\left|\sum \frac{X_i - m}{n}\right| \geqslant \varepsilon\right),$$

and the Chebyshev inequality gives

$$P\left(\left|\sum \frac{X_i - m}{n}\right| \geqslant \varepsilon\right) \leqslant \left\langle\left[\sum \frac{X_i - m}{n}\right]^2\right\rangle \Big/ \varepsilon^2$$

Inserting the variance of X, defined by

$$s^2 = \sum \frac{(X_i - m)^2}{n},$$

in the right-hand side and chaining the inequality, we have

$$P\left(\left|\frac{S_n}{n} - m\right| \geqslant \varepsilon\right) \leqslant \frac{s^2}{n\varepsilon^2}. \tag{A2.17}$$

As $n \to \infty$, this vanishes, proving the law of large numbers.

An example of the law of large numbers is provided by the discussion of the stochastic horse race in Section 4.4, where it was shown that the relative spread of the horses decreases as the length of the race increases. That is, if the horses' positions are plotted using the mean distance N/M as a unit, the horses will appear more tightly bunched as N increases.

The law of large numbers is often misinterpreted to imply that a losing streak will soon be compensated by a winning streak. This attitude is a sure way to disaster and it ignores the fact that past performance cannot influence a random process.

The law of large numbers does not operate by compensating past behavior; it operates by swamping the past.

APPENDIX 3

USE OF SPREADSHEETS
FOR SIMULATIONS

A3.1 WHY USE A SPREADSHEET?

Most of the simulation results demonstrated in this book were derived from conventional computer programs written in Basic or Pascal. Spreadsheet programs provide a powerful and easy alternative for some types of simulation. One advantage is that they require no programming, in the usual sense, and an inexperienced user can pick up enough skill in a few hours to construct useful spreadsheets. Then it becomes possible to set up and solve a least squares problem, for example, in a matter of minutes, whereas it usually requires the better part of an hour to get a simple program running from scratch when it is written in Pascal or Basic.

Virtually all spreadsheet programs have auxiliary graphing programs that work directly from the tables and produce instant graphs. So, in addition to their speed in setting up a problem, spreadsheets also permit seeing the solution at a glance, as a table or a graph. New solutions corresponding to new input data can be obtained as quickely as the data can be entered.

Thus, for the right type of problem, spreadsheets recommend themselves to the novice and to the expert.

A3.2 A FEW PROPERTIES OF SPREADSHEETS

This is not the place to explain the use of spreadsheets in detail. The aim of this chapter is merely to illustrate a few standard uses and a few tricks that can, with some imagination, make spreadsheets useful for simulations. The following examples are from Microsoft EXCEL programs.

A spreadsheet can be visualized as a rectangular array of cells; each cell contains an expression, which may be a numerical constant or a formula, whose numerical value is displayed. The formula is evaluated by referring to the numerical contents of other cells and performing prescribed calculations on them.

The task of setting up a spreadsheet consists of filling the required cells with the appropriate expressions. The calculation is then initiated by the user, and it is executed sequentially through the array. Automatic iteration is possible, so the calculation can be redone any number of times without intervention by the user.

Cells may refer to themselves. For example, the formula in a cell may say, "compute the new value of this cell by adding 1 to the old value." During each iteration, the numerical value of that cell will be incremented, and it can serve as an iteration counter.

Formulas or numbers can be copied from a cell to an entire row or column of cells, or from one block of cells to another block of the same shape, so it is an easy matter to set up large arrays.

One important limitation of spreadsheets will prevent their application to certain types of problem unless the user can find an ingenious way to get around it. That is, cells cannot compute the names of other cells. A cell must refer to predefined cells by name. To put it another way, it is not possible, as far as this user is aware, to have a cell refer to a cell whose name is computed.

We now provide a few examples.

A3.3 SOLUTION OF A LEAST SQUARES PROBLEM

We start with this appplication because it does not require iteration and it is closest to the original intended use of spreadsheets (which were evidently invented for accounting and "what–if" calculations). As an example, suppose we have five values of the measured distance of an object moving in a straight line. Referring to Table A3.1, the times t are in column B and the measured distances s are in Column C. We wish to fit a quadratic formula, $s = s_0 + vt + gt^2/2$, to the data by solving the normal equations, as discussed in Chapter 17. Successive columns contain formulas that give the computed values of t^2, t^3, t^4, st, and st^2. Their sums are found in row 8. For example, the formula in row 9 column D calls for the sum of the numbers in rows 4–8 in that column. Finally, the algebra leading to the coefficients v and g are entered in column B, rows 11 and 12.

When the measured data and the formulas have been entered, the computer is told to "Calculate Now," and the results appear almost instantly in the successive columns and then in column B, rows 11 and 12.

TABLE A3.1 Least Squares Spreadsheet Showing Solution to a Problem with Five Measured Distances _s_ at Times t^a

	A	B	C	D	E	F	G	H	I	J	K
3	i	t	s	tt	ttt	tttt	st	stt	y	e	ee
4	1	1	5	1	1	1	5	5	5.0803	-0.08	0.006
5	2	1.5	4	2.25	3.375	5.0625	6	9	3.9697	0.0303	9E-04
6	3	3	-14	9	27	81	-42	-126	-13.97	-0.034	0.001
7	4	4	-38	16	64	256	-152	-608	-38.09	0.092	0.008
8	5	5	-72	25	125	625	-360	-1800	-71.95	-0.046	0.002
9	sums			53.25	220.38	968.06	-543	-2520		-0.038	0.019
10											
11	v	9.9481									
12	g	9.7356									
13	sigma	0.0619									

^{a}The resulting initial speed v and acceleration are shown at the bottom with the standard deviation of the discrepancy between the formula and the data. Each cell evaluates a formula defining one aspect of the solution.

A3.4 RANDOM ARRAY

Most repreadsheet programs permit the calculation of a random variable, such as RND( ), that is uniformly distributed in the interval 0 to 1.0. Then, if each cell of a rectangular array is given by the formula **(IF (RND < 0.5),0,1)**, the computer interprets this to imply that the number in each cell is set to 0 if **RND < 0.5** is true; otherwise it is set to 1. When the computer calculates the array, the array is filled with a random field of 0s and 1s.

A3.5 EXPECTED ENDPOINTS OF ONE-DIMENSIONAL RANDOM WALK

As we saw in Chapter 3, the Pascal triangle describes the relative numbers of paths to various points in a discrete random walk on a line. A spreadsheet

TABLE A3.2 Random Walk with $p = 0.9$ of Stepping Left^a

	1	2	3	4	5	6	7	8	9	10	11	12	13	14	15	16	17	18	19
1	Walk	without	barriers																
2	PL	0.9	*=prob	of	stepping	left													
3										100									
4									90		10								
5								81		18		1							
6							73		24		2.7		0.1						
7						66		29		4.9		0.4		0					
8					59		33		7.3		0.8		0		0				
9				53		35		9.8		1.5		0.1		0		0			
10			48		37		12		2.3		0.3		0		0		0		
11			43		38		15		3.3		0.5		0		0		0		0

^{a}See Table A3.3 for a portion of the formulas.

TABLE A3.3   Formulas in a Portion of the Spreadsheet for the Random Walk Displayed in Table A3.2

	10	11
1		
2		
3		100
4	=R2C2*R[-1]C[1]	
5		=R[-1]C[-1]*(1-R2C2)+R[-1]C[1]*R2C2
6	=R[-1]C[-1]*(1-R2C2)+R[-1]C[1]*R2C2	
7		=R[-1]C[-1]*(1-R2C2)+R[-1]C[1]*R2C2
8	=R[-1]C[-1]*(1-R2C2)+R[-1]C[1]*R2C2	
9		=R[-1]C[-1]*(1-R2C2)+R[-1]C[1]*R2C2
10	=R[-1]C[-1]*(1-R2C2)+R[-1]C[1]*R2C2	
11		=R[-1]C[-1]*(1-R2C2)+R[-1]C[1]*R2C2
12		
13		

program can easily compute the entries of the Pascal triangle. Table A3.2 shows the solution to an asymmetric random walk in which 100 particles move with probability $p = 0.9$ to the left and $p(1 - p) = 0.1$ to the right at each step. Time increases downward, and each cell displays the number of particles that are expected to visit that site if 100 started. Column 20 shows the sum of the numbers of walkers in the corresponding rows, and they are constant, as no walkers are lost.

Table A3.3 shows the formulas in a small portion of the array. They say that each cell is the sum of two values: p times the value above and to the left plus

TABLE A3.4   Random Walk Like That of Table A3.3 with Absorbing Barrier (Upper) at the Line and a Reflecting Barrier (Lower)

	1	2	3	4	5	6	7	8	9	10	11	12	13	14	15	16	17	18	19
13	absorbing	barrier																	
14											100								
15										90		10							
16									81		18		1						
17								0		24		2.7		0.1					
18							0		22		4.9		0.4		0				
19						0		0		6.6		0.8		0		0			
20					0		0		5.9		1.4		0.1		0		0		
21				0		0		0		1.8		0.2		0		0		0	
22			0		0		0		1.7		0.4		0		0		0		0
23																			
24	reflecting	barrier																	
25																			
26																			
27											100								
28										90		10							
29									81		18		1						
30								73		24		2.7		0.1					
31							0		95		4.9		0.4		0				
32						0		85		14		0.8		0		0			
33					0		0		98		2.1		0.1		0		0		
34				0		0		88		12		0.3		0		0		0	
35			0		0		0		98		1.5		0		0		0		0

$1 - p$ times the value above and to the right. (The notation C2 means column 2 and C[1] means one column to the right.) The value of p is read from cell R2C2, so if that number is changed by the user, the entire table is recalculated and displayed at once.

Table A3.4 shows two variations, in which a barrier is inserted, as discussed in Chapter 6. The upper portion corresponds to an absorbing barrier, and it was achieved by merely setting the numerical values in column 8 equal to 0. Note that in this case the entries in column 20 no longer sum to 100, as some of the walkers are lost at the absorbing barrier. The lower portion of the figure shows the effect of a reflecting barrier achieved by setting the cells in column 7 equal to 0 and setting the formula in column 9 so the entry equals unity times the value in $R[-1]C[-1]$ plus $1 - p$ times the value in $R[-1]C[1]$. No walkers are lost at a reflecting boundary, so the entries in column 20 remain equal to 100.

A3.6   TEST OF A RANDOM-NUMBER GENERATOR (RNG)

The Wichmann–Hill RNG described in Chapter 1 has been used in Table A3.5 to fill the 10 bins (0–0.1, 0.1–0.2, etc.) listed in row 15. The upper diagram shows the spreadsheet before the calculation started, and the lower portion is the result of computing 500 numbers with the Wichmann–Hill RNG. The resulting bin counts are listed in row 18, and the uniformity of these numbers is a test of the RNG. Rows 3–10 contain the steps along the way to the random numbers, *RAND*, one of which is seen in R10C2. The formulas for this spreadsheet are displayed in Table A3.6.

TABLE A3.5   Layout for Testing of Random-Number Generator Described in Chapter 1

	1	2	3	4	5	6	7	8	9	10	11
1	Wichmann and Hill,		BYTE March 1987								
2											
3	21513	16174	16174								
4											
5	2874	9416	9416								
6											
7	24633	3036	3036								
8											
9	temp	1.6179104									
10	rand	0.6179104									
11	Running	1									
12	count	500									
13	expectation	50	sqrt(2n)=	4.472	n=	10					
14	Sum	253.74111	mean	0.507							
15	Bins	0.1	0.2	0.3	0.4	0.5	0.6	0.7	0.8	0.9	1
16											
17	Increment	0	0	0	0	0	0	1	0	0	0
18	Count	42	42	60	50	48	58	52	52	49	47
19	Sq Dev	1.28	1.28	2	0	0.08	1.28	0.08	0.08	0.02	0.18
20	Chi Sq	0.8318426									

TABLE A3.6 Formulas Used to Evaluate the Numerical Entries in Each Cell of Table A3.6

	1	2	3
1	Wichmann and Hill,		
2			BYTE March 1987
3	=RC[2]	=171*MOD(RC[-1],177)-2*INT(RC[-1]/177)	=IF(RC[-1]<0,RC[-1]+30269,RC[-1])
4			
5	=RC[2]	=172*MOD(RC[-1],176)-35*INT(RC[-1]/176)	=IF(RC[-1]<0,RC[-1]+30307,RC[-1])
6			
7	=RC[2]	=170*MOD(RC[-1],178)-63*INT(RC[-1]/178)	=IF(RC[-1]<0,RC[-1]+30323,RC[-1])
8			
9	temp	=R[-6]C[-1]/30269+R[-4]C[-1]/30307+R[-2]C[-1]/30323	
10	rand	=R[-1]C-INT(R[-1]C)	
11	Running	1	
12	count	=IF(R[-1]C>0,RC+1,0)	
13	expectation	=R12C2/R13C6	sqrt(2n)=
14	Sum	=IF(R11C2>0,RC+R10C2,0)	mean
15	Bins	0.1	0.2
16			
17	Increment	=IF(AND(R11C2=1,R10C2<0.1),1,0)	=IF(AND(R10C2>0.1,R10C2<0.2),1,0)
18	Count	=IF(R11C2=1,RC+R[-1]C,0)	=IF(R11C2=1,RC+R[-1]C,0)
19	Sq Dev	=IF(R13C2>0,(R[-1]C-R13C2)*(R[-1]C-R13C2)/R13C2,0)	=IF(R13C2>0,(R[-1]C-R13C2)*(R[-1]C-R13C2)/R13C2,0)
20	Chi Sq	=IF(R12C2>0,ABS((SUM(R[-1]C:R[-1]C[9])-R13C6/R13C4),0)	

One important feature of this example is the use of an index, *Running*, in R11C2. This number is set by the user. When *Running* = 0, the bin counts are reset to 0; when *Running* = 1, the iterations proceed and the counts are accumulated. Note that in the formulas for row 18, the value of *Running* is tested and used to decide whether to reset to 0 or to accumulate by adding in the value above, which is the *increment*. The entry in R12C2 is the counting index.

REFERENCES

Atkins, P. W., 1984. *The Second Law.* The Scientific American Library. New York: W. H. Freeman.

Baker, G., 1986. "A Simple Model of Irreversibility." *Am. J. Phys.*, **54**(8):704–708. *(One-dimensional gas of hard atoms.)*

Bennett, W. R. Jr., 1976. *Scientific and Engineering Problem-Solving with the Computer.* Englewood Cliffs, NJ: Prentice-Hall. *(Introduction to computational aspects.)*

Binder, K., ed., 1984. *Application of the Monte Carlo Method in Statistical Physics.* New York: Springer-Verlag. *(Comprehensive reviews of recent research.)*

Boyd, Ian D., 1989. "Monte Carlo Simulations of an Expanding Gas." *Comput. Phys.*, **May/June**:73–76. *(Contains several useful references.)*

Brush, S., 1983. *Statistical Physics and the Atomic Theory of Matter: From Boyle and Newton to Landau and Onsager.* Princeton: Princeton University Press. *(Historical.)*

Chandrasekhar, S., 1943. "Stochastic Problems in Physics and Astronomy." *Rev. Mod. Phys.* **15**:1. *(A classic, although not for the novice.)*

Chowdhury, D., and A. Mookerjee, 1985. "Random Walk and Magnetization of Spin Clusters in Spin Glasses." *Am. J. Phys.*, **53**(3):261–263.

Ciccotti, G., D. Frenkel, and I. McDonald, 1987. *Simulations of Liquids and Solids: Molecular Dynamics and Monte Carlo Methods in Statistical Mechanics.* Amsterdam: North-Holland. *(Annotated reprints of several dozen key papers from American literature since 1953.)*

Cleveland, B. T., 1983. "The Analysis of Radioactive Decay with a Small Number of Counts by the Method of Maximum Likelihood." *Nucl. Instrum. Methods*, **214**:451. *(Explanation and examples of maximum likelihood for cases when least squares will not work well.)*

Corey, Ellen M., and David A. Young, 1989. "Optimization of Physical Data Tables by Simulated Annealing." *Comput. Phys.*, **May/June**:33–37.

Crawford, F., 1988. "Using Einstein's Method to Determine Both the Planck and Fermi–Dirac Distributions." *Am. J. Phys.*, **56**(10):883–885.

Dempsey, D., and J. Hartman, 1986. "*PV* Cycle Area Equals Work Done: An Undergraduate Experiment." *Am. J. Phys.*, **54**(12):1086–1088.

Efron, B., and C. Morris, 1977. "Stein's Paradox in Statistics." *Sci. Am.*, **May**:119–127. (*When the average is not the best estimator of future behavior.*)

Ehrlich, P., 1981. "The Concept of Temperature and Its Dependence on the Laws of Thermodynamics," *Am. J. Phys.*, **49**(7):622–631. (*Discusses the basis for formal definition of temperature.*)

Eigen, M., and R. Winkler, 1983. *Laws of the Game.* New York: Harper & Row. (*Excellent discussion of board games based on chance.*)

Feder, Jens, 1988. *Fractals: Physics of Solids and Liquids.* New York: Plenum Press. (*Comprehensive elementary introduction to research applications of fractal geometry, including percolation and random walks.*)

Feller, W., 1968. *An Introduction to Probability Theory and Its Applications*, 3rd ed. New York: Wiley (*A gem, although too advanced for most college freshmen.*)

Feynman, R. P., 1972. *Statistical Mechanics. A Set of Lectures.* Edited by David Pines. Frontiers in Physics. Menlo Park, CA: Benjamin-Cummings.

Gardner, M., 1959. "Mathematical Games: Problems Involving Questions of Probability and Ambiguity." *Sci. Am.*, **October**:174–182. (*Problems in continuous sample spaces.*)

Gould, H., and J. Tobochnik, 1988. *An Introduction to Computer Simulation Methods*, 2 vols. Reading, MA: Addison-Wesley. (*Excellent and practical computational introduction to a variety of topics, with good reference lists at the end of each chapter.*)

Gurney, R. W., 1949. *Introduction to Statistical Mechanics.* New York: McGraw-Hill. (*An introduction to the Boltzmann distribution that strongly influenced the approach used in this book.*)

Hersch, R., and R. Griego, 1969. "Brownian Motion and Potential Theory." *Sci. Am.*, **March**:67–74. (*Introduction to use of random walk to solve differential equations.*)

Hill, T. L., 1963. *Thermodynamics of Small Systems.* Elmsford, NY: Benjamin.

Hoel, P. G., S. C. Port, and C. J. Stone, 1971. *Introduction to Probability Theory.* Boston: Houghton Mifflin. (*One of my favorites.*)

Hoel, P. G., S. C. Port, and C. J. Stone, 1972. *Introduction to Stochastic Processes.* Boston: Houghton Mifflin. (*Treats topics slightly more advanced mathematically than in the present book.*)

Kalos, Malvin H., and Paula A. Whitlock, 1986. *Monte Carlo Methods. I. Basics.* New York: Wiley. (*Systematic elementary introduction to the mathematical methods with emphasis on random walks.*)

Kerrich, J. E., 1946. *An Experimental Introduction to the Theory of Probability.* Copenhagen: E. Munksgaard.

Kinsella, A., 1986. "Numerical Methods for Error Evaluation." *Am. J. Phys.*, **54**(5):464–466. (*Examples of jackknife and bootstrap methods of error estimation.*)

Kirkpatrick, S., C. D. Gellat Jr., and M. P. Vecchi, 1983. "Optimization by Simulated Annealing." *Science*, **220**:671–680. (*The best introduction to simulated annealing and some of its applications.*)

Kittel, C., and H. Kroemer, 1980. *Thermal Physics*. New York: W. H. Freeman (*Very thorough, physically oriented introduction requiring calculus.*)

Knuth, D. E., 1981. *The Art of Computer Programming*, 2nd ed. Reading, MA: Addison-Wesley. (*Tough going, but a brilliant discussion of random number testing, among many topics.*)

Koonin, S., 1986. *Computational Physics*. Menlo Park, CA: Benjamin-Cummings. (*Undergraduate treatment.*)

Kruglak, H., 1987. "Brownian Movement: An Improved TV Demonstration." *Am. J. Phys.*, **55**(10):955–956. (*Ingenious method of demonstrating Brownian motion.*)

Metropolis, N., A. Rosenbluth, M. Rosenbluth, A. Teller, and E. Teller, 1953. "Equation of State Calculations by Fast Computing Machines." *J. Chem. Phys.*, **21**:1087–1092. (*The seminal paper on importance sampling and simulated annealing.*)

Miller, A. R., 1981. *BASIC Programs for Scientists and Engineers*. Berkeley: Sybex. (*Listings of many useful programs in BASIC.*)

Mishima, N., T. Y. Petrosky, H. Minowa, and S. Goto, 1980. "Model Experiment of Two-Dimensional Brownian Motion by Microcomputer." *Am. J. Phys.*, **48**(12):1050–1055. (*Assumes random scattering by large Brownian particle; includes flowchart and BASIC listing.*)

Mohling, F., 1982. *Statistical Mechanics: Methods and Applications*. New York: Wiley.

Monod, J., 1972. *Chance and Necessity*. New York: Random House. (*Microbiologist views randomness in life and discusses the emergence of form. A popular treatment, but demanding.*)

Mosteller, F., R. E. K. Rourke, and G. B. Thomas Jr., 1970. *Probability with Statistical Applications*. Boston: Addison-Wesley. (*One of the clearest introductions for college undergraduates.*)

Papoulis, Athanasios, 1965. *Probability, Random Variables, and Stochastic Processes*. McGraw-Hill Series in Systems Science. New York: McGraw-Hill. (*Comprehensive introduction to probability and stochastic processes at an advanced undergraduate level.*)

Prigogine, I., 1978. "Time, Structure, and Fluctuations." *Science*, **201** (1 September): 777–785. (*Nobel prize talk on the foundations of nonequilibrium thermodynamics.*)

Rechtman, R., 1988. "An Adiabatic Reversible Process." *Am. J. Phys.*, **56**(12):1104–1105.

Reichl, L. E., 1980. *A Modern Course in Statistical Physics*. Austin: University of Texas Press. (*Comprehensive introduction to the mathematical formalisms of statistical thermodynamics.*)

Reif, F., 1965. *Fundamentals of Statistical and Thermal Physics*. New York: McGraw-Hill. (*My favorite college text on the subject.*)

Reif, F., 1967. *Statistical Physics*. Berkeley Physics Course. New York: McGraw-Hill.

Rio, Fernando del, Maria Trigueros, and Esteban Martina, 1976. "On the Generalized Bernouli Construction. II. Extension to Two and Three Dimensions." *Am. J. Phys.*, **44**(1):36–40.

Ripley, Brian D., 1987. *Stochastic Simulation*. Wiley Series in Probability and Mathematical Statistics. New York: Wiley (*Intended for statisticians and researchers using simulation techniques; discusses variance reduction and uses of simulation.*)

Rubinstein, Reuven Y., 1981. *Simulation and the Monte Carlo Method*. Wiley Series in Probability and Mathematical Statistics. New York. Wiley. (*Introductory treatment with good discussion of random numbers and generation of random variates.*)

Schumacher, R., 1986. "Brownian Motion by Light Scattering Revisited." *Am. J. Phys.*, **54**(2):137–141.

Sherwood, B., and W. Bernard, 1984. "Work and Heat Transfer in the Presence of Sliding Friction." *Am. J. Phys.*, **51**(11):1001–1007.

Stanley, H. E., and N. Ostrowsky, 1986. *On Growth and Form: Fractal and Non-Fractal Patterns in Physics.* Dordrecht: Martinus Nijhoff. (*Lecture notes assuming some familiarity with the topic. Many interesting applications.*)

Stevens, P., 1974. *Patterns in Nature.* Boston: Little, Brown & Co. (*Largely qualitative discussion of mathematics of natural forms. Excellent illustrations.*)

van Ness, H. C., 1983. *Understanding Thermodynamics.* New York: Dover. (*Intended as supplement to undergraduate text.*)

Vincenti, W. G., and C. H. Kruger Jr., 1965. *Introduction to Physical Gas Dynamics.* New York: Wiley. (*A pioneering book that remains an excellent introduction.*)

Wax, N., 1954. *Selected Papers on Noise and Stochastic Processes.* New York: Dover. (*A collection of important modern presentations.*)

Weaver, W., 1982. *Lady Luck.* New York: Doubleday. (*Excellent introduction to probabilistic games and their history.*)

Wheeler, J. A., 1983. *Am. J. Phys.*, **51**:398. (*Applications to physics.*)

Whitney, C., 1984. "Departures from Thermal Equilibrium in Expanding Stars." *Astrophys. J.*, **278**:310–317. (*Mathematical discussion of stars as open systems.*)

Whitney, C., 1986. "Casino Physics in the Classroom." *Am. J. Phys.*, **54**(12):1079–1085. (*Description of simulations for college students.*)

Whitney, C., 1989. "Random Searches." In *Statistics: A Guide to the Unknown*, edited by J. Tanur, San Francisco: Holden-Day. (*Nontechnical introduction to optimization.*)

Wichmann, B., and D. Hill, 1987. "Building a Random-Number Generator." *Byte* **March**:127–128. (*Provides a Pascal program for random numbers.*)

Yourgrau, W., A. van der Merwe, and G. Raw, 1982. *Treatise on Irreversible and Statistical Physics.* New York: Dover. (*An excellent advanced discussion of modern problems of thermodynamics.*)

INDEX